International Conference on Forest Products (ICFP 2018)

Efficiency and Quality Improvement of Forest Products Towards Technology 4.0

IOP Conference Series: Earth and Environmental Science
Volume 359

Bogor, Indonesia
2 November 2018

ISBN: 978-1-7138-0155-9
ISSN: 1755-1307

Printed from e-media with permission by:

Curran Associates, Inc.
57 Morehouse Lane
Red Hook, NY 12571

Some format issues inherent in the e-media version may also appear in this print version.

Institute of Physics
Dirac House, Temple Back
Bristol BS1 6BE UK

Phone: 44 1 17 929 7481
Fax: 44 1 17 920 0979

techtracking@iop.org

Additional copies of this publication are available from:

Curran Associates, Inc.
57 Morehouse Lane
Red Hook, NY 12571 USA
Phone: 845-758-0400
Fax: 845-758-2633
Email: curran@proceedings.com
Web: www.proceedings.com

TABLE OF CONTENTS

IOP Conf. Series: Earth and Environmental Science **359** (2019) 011001 doi:10.1088/1755-1315/359/1/011001

The International Conference on Forest Product (ICFP) 2018 is held based on the consideration to broaden the forest product science and technology dissemination. As we all know, the dissemination is an integral part of the research activity which is used to spread the knowledge and inform the result to all stakeholders to ensure its expedience.

The ICFP 2018 is also an implementation of the role of the Forest Products Research and Development Center as the center of excellence for tropical forest products science and technology, and as the center of excellence, we are demanded to have the capability to hold outstanding activities in international level. The ICFP 2018 is expected to be a medium for researcher, scientist, and all stakeholders from various research institute and universities to exchange best practices and information as well as to develop research networking.

The theme of the ICFP 2018 " Efficiency and Quality Improvement of Forest Products Towards Technology 4.0" is a strategic theme to boost the role of science and technology for sustainable management of forest resources and environment. The proceeding presented 14 results of research on forest harvesting, bio-composite technology, non-timber forest products, basic properties of wood, and wood drying. I hope the results of research that presented in this proceeding can be beneficial for the parties and become a benchmark for future research and development, in Indonesia and at the international.

Thank you.

Dr. Wening Sri Wulandari
Chairman of Organizing Committee

Peer review statement

All papers published in this volume of *IOP Conference Series: Earth and Environmental Science* have been peer reviewed through processes administered by the proceedings Editors. Reviews were conducted by expert referees to the professional and scientific standards expected of a proceedings journal published by IOP Publishing.

Use of Organic Resins for Wood Modification

J. Balfas

Forest Product Research and Development Center,
Research, Development and Innovation Agency (FORDIA),
Ministry of Environment and Forestry, Bogor, Indonesia

email: jamalbs2000@yahoo.com

Abstract. Several Indonesian wood species were modified with two solutions of organic resins, i.e. shellac and damar which were diluted with methanol. The resin solutions were impregnated into wood blocks of coconut, rubber, fast grown teak and jabon. Physical and mechanical properties, and durability of the treated woods were determined to evaluate effects of the treatments. Results showed that effects of the treatments varied depending on wood species and kind of treatment. Wood blocks, treated to the highest weight percent gain (WPG), attained approximately 15% anti-shrink efficiency (ASE) after ten cycles of water saturation and drying. Coconut wood has higher permeability, solution uptake, retention and WPC than those of other three wood species. The ASE increased with an increase in the WPG. The resin modifications induced significant durability and mechanical strength improvements on all wood species.

1. Introduction

Wood is an important and favorable building material because it is economical, low in processing energy, renewable, strong, and aesthetically pleasing. On the contrary, there are several wood disadvantageous properties such as biodegradability, flammability, changing dimensions with varying moisture contents, and degradability by ultraviolet (UV) light, acids, and bases. Such inferior properties of wood have long been subjects of improvement through various modification treatments.

Wood characteristic improvements can be done through physical, mechanical or chemical treatments, but the last is most intensively explored [1, 2, 3, 4, 5]. Hill [2] also explained that the modification of wood can involve active modifications, which result in a change to the chemical nature of the material, or passive modifications, where a change in properties is effected, but without an alteration of the chemistry of the material. The active modification methods investigated to date have involved the chemical reaction of a reagent with the cell wall polymer hydroxyl groups. This can result in the formation of a single chemical bond with one OH group, or cross-linking between two or more OH groups.

There have been numerous efforts undertaken to improve characteristics of Indonesian woods using imported reagents, monomers and polymers. Balfas [6] reported that furfurylation treatment significantly improve dimensional stability and strength of tusam (*Pinus merkusii*) and mangium (*Acacia mangium*). Wood modification on jabon (*Anthocephalus* sp.) and nyamplung (*Callophyllum* sp.) with acetylation markedly improved vapor arbsortion property of the two wood species [7]. Significant improvements were demonstrated through the impregnation of the synthetic monomers and polymers into jabon wood [8]. However, these imported impregnants are very expensive and hardly meet economic production requirements.

Alternative approach has to be sought to find an effective way for improving wood quality and economically feasible as well. The use of certain wood extractive and organic resins to improve other wood characteristics has been initiated by Yamamoto and Hong [9]. Extractive impregnation has significant improvements on physical and biological wood performances [10]. Pine resin impregnation could improve strength and machining properties of eight wood species [11].

It is established that wood modification with organic resin is able to influence dimensional stability, strength and durability of the treated wood [12, 13]. It can be an alternative to conventional preservation of wood and enhances the resistance to biological destruction agents without using biocidal compounds. Treatment of wood with organic resin is classified as a passive (impregnation) modification, where the resin is deposited in the cell wall without chemically reacting with the matrix polymers [14].

This study aims at investigating the effect of organic resin solutions impregnation to enhance dimensional stability, strength and durability of the modified wood.

2. Material and methods

2.1. Wood and chemicals

Wood logs of coconut (*Cocos nucifera* L.), rubber (*Hevea brasiliensis* Mull.Arg.), fast grown teak (*Tectona grandis* L.f.) and jabon (*Anthocephalus cadamba* Roxb. Miq.) were collected from a farmer plantation in west Java and cut into planks of approximately 60×150×2000 mm^3 (radial × tangential × longitudinal) or (tangential × radial × longitudinal). The planks were cut to the specific grain orientation and specimen sizes described below. The specimens were then oven-dried at 65 ± 2°C for several days to reach approximately 10% moisture content. Samples were then weighed with electric balance in order to be able to determine the solution uptake and weight percent gain (WPG) after modification. Prior to treatment with organic resins, the dried wood specimens were conditioned at 25°C and 65% relative humidity (RH) to a constant weight.

2.2. Organic resins treatment

Resin solution formulations were prepared with 4 and 8% weight of shellac and damar, respectively, by diluting with technical grade methanol. The conditioned wood specimens were impregnated at 25°C in a vacuum-pressure vessel using a full cell process which included an initial vacuum phase of 5 kPa (15 minutes) and a pressure phase of 1200 kPa (1 hour). After the impregnation, the excess solution at the surface of the samples was removed with a paper towel and the samples were weighed. Then they were exposed to the temperature of 65°C in a drying oven for 48 hours. Solution uptakes (SU) after treatment and the weight percent gain (WPG) after drying were related to the dry mass of the untreated wood samples.

2.3. Anti-shrink efficiency and maximum swelling determination

Anti-shrink efficiency (ASE) was determined after cyclic water soaking and drying as described by Xiao et al. [15]. Five specimens (25 mm × 25 mm × 10 mm, r × t × l) per treatment and 5 untreated specimens were subjected to 10 cycles of water saturation (1 h with distilled water at 8 kPa, 24 h stored under water at atmospheric pressure) subsequently oven drying (40°C, 80°C, 103°C; 24 h each). After each cycle, the dimensions at oven dry state and water saturation were determined.

2.4. Durability

Wood durability testing against subterranean termites (*Coptotermes cynocephalus*) was conducted by the procedure of an Indonesian standard [16]. Five specimens (25 mm × 25 mm × 5 mm, r × t × l) for each treatment were placed in glass containers with 200 g of sterilized river sand, 50 mL of water and 200 healthy and active workers of *C. cynocephalus* termites from a laboratory colony. The containers were put in a dark room at temperature of 25–30°C and 80–90% relative humidity for 4 weeks and weighed each week. If the moisture content of the sand decreased by 2% or more, water was added to maintain the 25% moisture content. At the end of the fourth week, wood specimens were oven dried to determine weight loss, termite mortality and termite feeding rate, using the following formula.

$$\text{Weight loss} = (W_1 - W_2)/W_1 \times 100\% \quad (1)$$

where W1 = weight of the oven-dried sample before the test (g)
W2 = weight of the oven-dried sample after the test (g).

$$\text{Termite mortality} = (T_1 - T_2)/T_1 \times 100\% \quad (2)$$

where T1 = number of live termites before to the test and
T2 = number of live termites after the test.

The calculated results were then classified into durability class according to the following table.

Table 1. Resistance class against subterranean termite [16]

Resistance	Classification	Mass loss (%)
I	Very resistant	< 3.52
II	Resistant	3.52–7.50
III	Moderate	7.50–10.96
IV	Poor	10.96–18.94
V	Very poor	> 18.94

2.5. Mechanical properties
Parallel and perpendicular compression strengths of the treated and untreated samples were determined according to the modified ASTM standard [17]. Parallel compression strength specimens (20 mm × 20 mm × 90 mm, r × t × l) and perpendicular compression strength specimens (20 mm × 20 mm × 60 mm, r × t × l) were conditioned at 25°C and 65% relative humidity (RH) to a constant weight prior to mechanical testing.

3. Results and discussion

3.1. Weight percent gain and solution uptake.
Solution uptake (SU), retention, and weight percent gain (WPG) demonstrated that four wood species were easily treatable with two organic resin formulations (Table 2). Both parameters linearly increased with the solid content in the treating solution. As observed previously [12], the weight-based solution uptake (SU) tended to slightly increase with the solid content as the density of the treating solution increases with the solid content. Table 2 showed that coconut wood samples have higher SU, retention and WPG than those of the other species. This may be due to the more porous structure of coconut wood with abundant parenchymatous tissue and metaxylem. The fast grown teak (FGT) samples indicated lower values of SU, retention and WPG. The least permeability of fast grown teak (FGT) may be related to its smaller pith perforation and vessel size compared with the other treated wood samples [13]. Table 2 also showed that the samples treated with damar had higher SU, retention and WPG than those treated with shellac resin. This difference may be caused by the higher particulate content per unit volume of the damar solutions, as damar resin is known to possess higher specific gravity of about 1.4 [18] than 1.1 of shellac resin [19].

Table 2. Solution uptake, retention and WPG of the treated wood

Treatment	Solution uptake (%)				Retention (kg.m-3)				WPG (%)			
	Co	Rw	FGT	Jb	Co	Rw	FGT	Jb	Co	Rw	FGT	Jb
4% Shellac	107	88	81	92	134	112	108	116	57	23	18	30
8% Shellac	118	93	87	101	157	121	114	130	62	26	22	36
4% Damar	136	105	93	108	196	154	137	162	68	34	26	42
8% Damar	154	119	102	124	216	160	148	175	76	36	31	48

Remarks: Co=Coconut wood; Rw=Rubber wood; FGT=Fast grown teak; Jb=Jabon

3.2. Anti-shrink efficiency (ASE) and maximum swelling (SW_{max})

ASE and SW_{max} show a dimensional stability of the modified wood samples. Depending on the resin concentration and wood species as shown in Table 2, the modification with organic resins induced high bulking in the range of weight gains 22-76%. These high bulking values are attributed to the relatively light weight of the resin solutions (specific gravity below 0.85). In addition, drying prior to the modification might have reduced the diameter of cell wall pores as compared to the never dried wood [20].

Maximum swelling of coconut, rubber, FGT and jabon during ten saturation and drying cycles was respectively plotted in Figures 1-4. As a consequence of resin bulking, the modifications imparted a reduced SWmax and moderate ASE, which was continuously decreased by increasing the number of cycle as shown in Figures 5-8. Resin bulking and ASE increased with increasing WPG. With respect to the high WPG attained at high treatment concentration, however, bulking and ASE were relatively low compared to the active modifications such as acetylation. This indicated that major proportion of the organic resin was not accommodated in the cell wall.

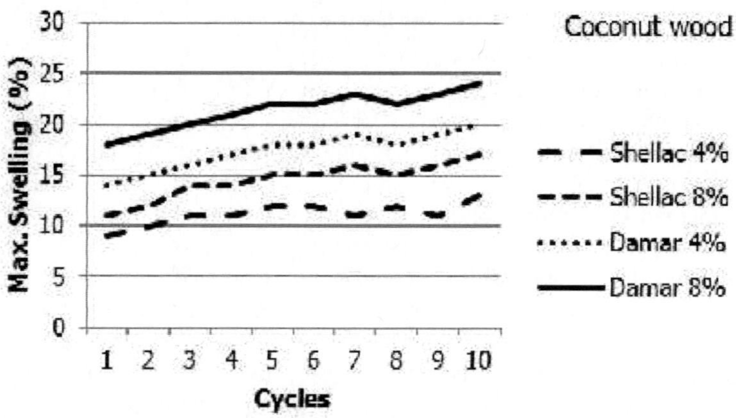

Figure 1. Maximum swelling of coconut wood over 10 cycles saturation and drying

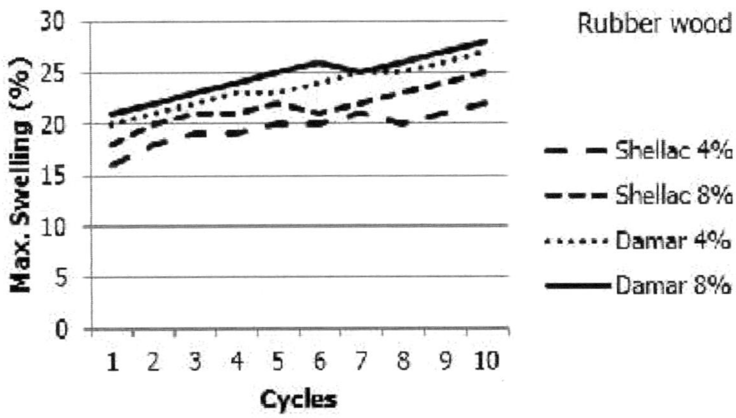

Figure 2. Maximum swelling of rubber wood over 10 cycles saturation and drying

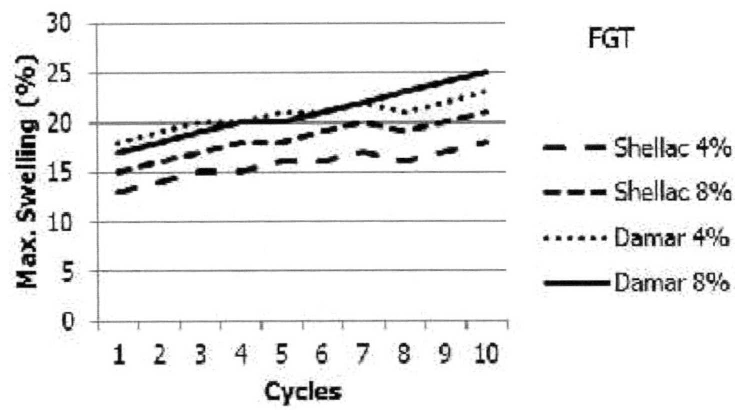

Figure 3. Maximum swelling of FGT over 10 cycles saturation and drying

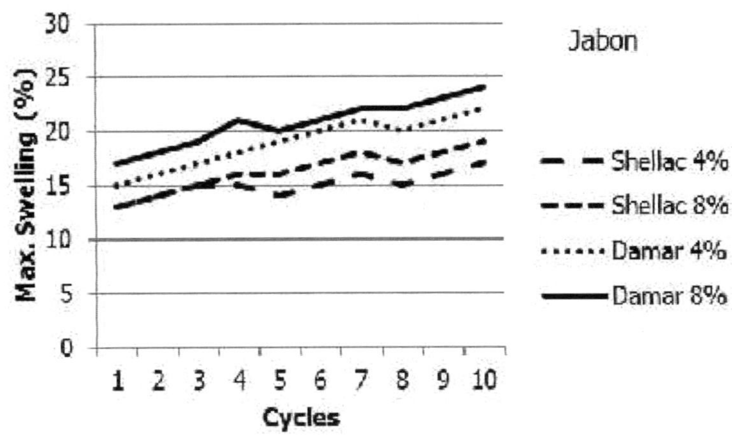

Figure 4. Maximum swelling of jabon over 10 cycles saturation and drying

The SWmax of the modified specimens (Figures 1-4) increased with an increase in the number of saturation and drying cycles. This is particularly attributed to leaching of hydrophilic extractives and to hornification of the cell wall due to high temperature drying [21]. The phenomena can be explained by bulking reduction due to leaching during cyclic wetting and drying and rearrangement of the resins in the cell wall. The decreasing ASE with increasing number of wetting and drying cycles is due to the effect of a slight increase in the SWmax values of the resin-modified specimens. This steady minor loss in the ASE indicates that the long-term outdoor dimensional stability of the organic resin-treated wood must be called into questions.

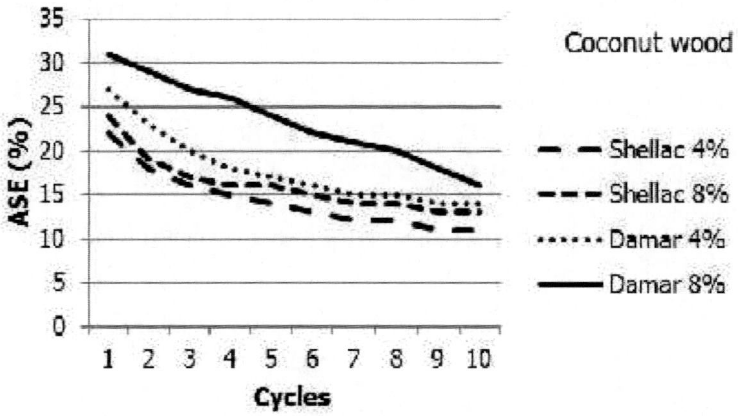

Figure 5. ASE of coconut wood over 10 cycles saturation and drying

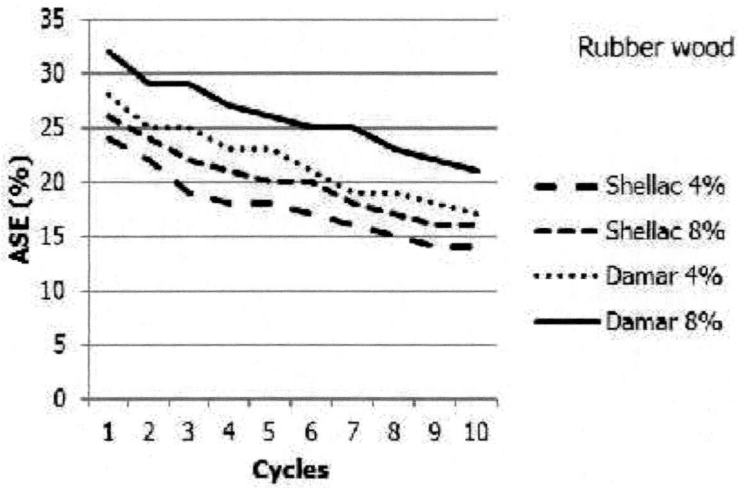

Figure 6. ASE of rubber wood over 10 cycles saturation and drying

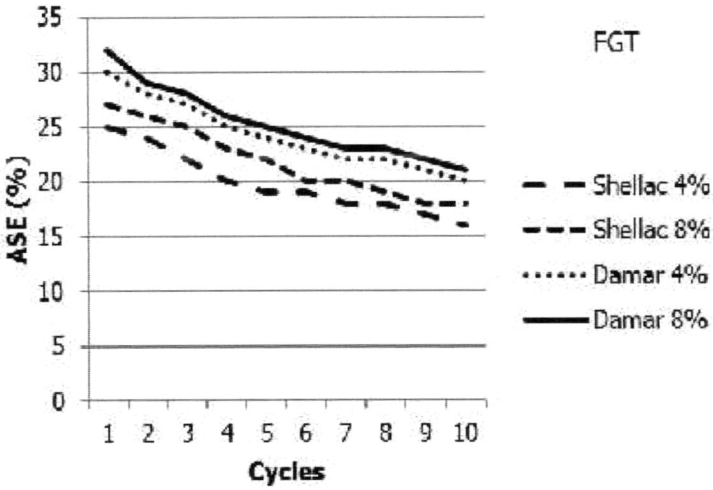

Figure 7. ASE of FGT over 10 cycles saturation and drying

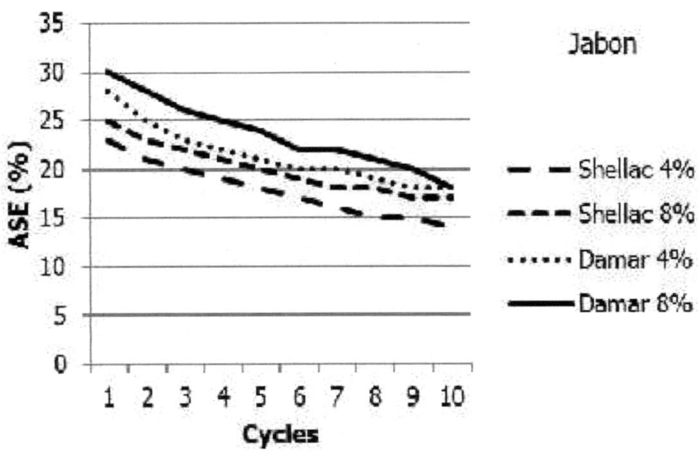

Figure 8. ASE of jabon over 10 cycles saturation and drying

3.3. Durability

The weight loss of wood sample, termite mortality and termite feeding rate after 4-week subterranean termite test are shown in Table 3. Values of each parameter are significantly different from treatments and wood species.

IOP Conf. Series: Earth and Environmental Science **359** (2019) 012001 doi:10.1088/1755-1315/359/1/012001

Table 3. Weight loss, mortality and feeding rate on termite test

Treatment	Weight loss (%)				Mortality (%)				Feeding rate (mg termite^{-1} week^{-1})			
	Co	Rw	FGT	Jb	Co	Rw	FGT	Jb	Co	Rw	FGT	Jb
Untreated	32.	28.3	28.5	30.8	42	48	46	45	65	61	60	63
4% Shellac	7.5	7.8	7.1	7.5	96	98	98	96	18.9	17.2	17.1	18.2
8% Shellac	6.8	7.3	6.7	7.3	98	98	99	97	18.5	16.8	16.0	17.6
4% Damar	5.8	6.5	6.3	6.8	99	99	99	97	18.3	16.2	15.2	16.3
8% Damar	5.7	5.9	5.1	6.4	99	99	99	98	16.2	14.1	13.8	16.0

Remarks: Co=Coconut wood; Rw=Rubber wood; FGT=Fast grown teak; Jb=Jabon

The weight loss of untreated wood samples varied from 28.3 to 32.6%, while the resin treated wood samples had weight loss range from 5.1 to 7.8%. The weight loss of all specimen was affected by wood species, resin concentration and the interaction of these factors. Impregnation FGT with 8% damar had the lowest weight loss compared with those of other wood species and resin treatment. The highest weight loss was encountered in rubber wood samples impregnated with 4% shellac resin. Table 3 clearly indicates that treatment with higher resin content resulted in lower wood weight loss. Table 3 also demonstrates that impregnation using damar resin consistently had lower weight losses than those treated with shellac resin. Results of this study are in line with those of Basri and Balfas [13] who stated that resin loading is proportionally increased with increasing resin content.

Based on weight losses, we can determine the wood resistance class according to the Indonesian standard [16]. The untreated wood samples, having the weight loss of 28.3 to 32.6% (Table 3), were all categorised into class V (Table 1), which is very poor resistance against subterranean termite attack. And most of the resin treated wood samples were classified into class II, i.e. resistant. This indicates that organic resin treatment is very effective in increasing wood resistance against termites. The efficacy of organic resins in improving durability of jabon wood is comparable to the durability improvement made on the same wood using methyl methacrylate [8].

Termite mortality is significantly varied according to wood species, the resin treatment, and interaction of the two factors. The average termite mortality in the resin treated wood samples (Table 3) varied from 96 to 99%, meaning that most all termites were dead.. These values are about double of the untreated samples (mortality of 42 to 48%). The organic resins were not toxic to termites, but they formed a barrier in the wood structure, preventing termites to reach the wood tissues directly. Higher resin deposition would result in stronger barrier and greater resistance. Resin treated wood samples are markedly more resistant than the untreated samples against *C. cynocephalus* subterranean termites.

Termite feeding rate on the untreated wood samples varied between 60 to 65 mg termite^{-1} week^{-1}. These values are lower than the standard feeding rate of pine (*Pinus merkusii*), i.e. 79 mg termite^{-1} week^{-1} [22]. The resin treated wood had much lower feeding rates of 14 to 19 mg termite^{-1} week^{-1}, which are about 30% of the untreated wood termite feeding rate. Wood species, impregnation treatment, and interaction of the two factors significantly affected values of the termite feeding rate. The lowest feeding rate was encountered in treatment fast grown teak with 8% damar.

3.4. Mechanical strength

Coconut, rubber, fast-grown teak and jabon wood samples treated with organic resins showed significant improvement in mechanical properties (Table 4). The effect varied depending on wood species, modification treatment, and combination of the two factors. Coconut wood samples treated with organic resins possess approximately 20 to 60% higher compression strength compared to those of the untreated samples (Table 4). The strength improvement of coconut wood is recorded significantly higher than those of the other three wood species, which increased of 10~30% compared to their untreated samples. The difference may be related to the higher solution uptake, retention and WPG of coconut wood compared with the other wood species (Table 2). Effective resin retention in the structure of wood samples may have caused wood structure to bulk and fibre bonding to increase to contribute to mechanical improvement [2]. Table 4 also shows that the modification treatment with damar resin improved mechanical properties of wood samples better than those treated with shellac resin.

Table 4. Parallel and perpendicular compression strength (kg.cm^{-2})

Treatment	Compress // grain				Compress ⊥ grain			
	Co	Rw	FGT	Jb	Co	Rw	FGT	Jb
Untreated	232.20	442.20	236.21	142.40	74.05	186.80	74.63	34.52
4% Shellac	266.31	458.90	252.08	152.51	96.24	242.51	75.92	42.38
8% Shellac	282.37	488.40	264.38	160.22	100.22	260.59	78.37	44.40
4% Damar	290.63	503.26	270.30	170.35	113.70	282.40	92.64	48.61
8% Damar	312.26	542.05	288.26	184.68	128.06	296.26	98.63	62.48

Remarks: Co=Coconut wood; Rw=Rubber wood; FGT=Fast grown teak; Jb=Jabon

4. Conclusions

Treatments of four wood species with two resin solutions of shellac and damar has resulted in various solution uptake, retention, weight percent gain, durability and mechanical strength between the wood species. Coconut wood has higher permeability, solution uptake, retention and weight percent gain than the other wood species. The anti-shrink efficiency (ASE) increased with an increase in the WPG. The resin modifications induced significant durability and mechanical strength improvements on all wood species.

References

[1] Rowell, R. M. 2005. *Handbook of Wood Chemistry and Wood Composites.* Taylor and Francis Group. CRC Press.

[2] Hill, C. A. S. 2006. *Wood Modification: Chemical, Thermal, and Other Processes* (School of). Bangor, England: John Wiley & Sons Ltd.

[3] Bowyer JL, Shmulsky R, Haygreen JG. 2007. *Forest Products & Wood Science: An Introduction.* (5th Edition). Ames: Iowa State Press.

[4] Pandey, K.K., Jayashree & Nagaveni, H.C. 2009. Study of dimensional stability, decay resistance, and light stability of phenylisothiocyanate modified rubberwood. Bioresources **4** (1): 257-267.

[5] Ibach, R. E. (2010). *Specialty Treatments. Wood Handbook Chapter 19.* Madison, Wisconsin: *Wood Research* Forest Products Laboratory, USDA.

[6] Balfas, J. 1995. Furfurilasi pada kayu tusam (Pinus merkusii Jungh.et de Vr.) dan mangium (Acacia mangium Willd.). *Jurnal Penelitian Hasil Hutan* **13**(5):176-185.

[7] Krisdianto. 2012. Pengaruh asetilasi terhadap penyerapan uap air pada dua jenis kayu tropis. *Jurnal Penelitian Hasil Hutan* **30**(2): 94-99.

[8] Hadi, Y.S., Rahayu, I.S. & Danu, S. 2015. Termite resistance of jabon wood impregnated with methyl methacrylate. *Journal of Tropical Forest Science*. **27**(1):25-29.

[9] Yamamoto, K. & Hong. L. T. (1988). Decay resistance of extractives from chengal (Neobalanocarpusheimii). *Journal of Tropical Forest Science*, *1*(1), 51–55.

[10] Kjellow, A.W. & Henriksen, O. 2009. Supercritical wood impregnation. *The Journal of Supercritical Fluids* 50(3): 297-304.

[11] Balfas, J. 2000. Penyempurnaan kayu dengan perlakuan modifikasi JRP-2. Prosiding Lokakarya Penelitian Hasil Hutan. Bogor: Pusat Penelitian Hasil Hutan.

[12] Balfas, J. (2007). Perlakuan resin pada kayu kelapa (Cocos nucifera). *Jurnal Penelitian Hasil Hutan*, ***25***(*2*), 108–118.

[13] Basri, E.dan J. Balfas. (2014). Impregnasi ekstrak jati dan resin pada kayu jati cepat tumbuh dan karet. *Jurnal Penelitian Hasil Hutan, **32** (4),*283-289.

[14] Norimoto, M., Gril, J. and Rowell, R.M. (1992). Rheological creep properties of chemically modified wood. Relationship between dimensional and creep stability. *Wood and Fiber Science*, **24**(10), 25–35.

[15] Xiao, Z., Xie, Y., Militz, H., Mai, C., 2010. Effect of glutaraldehyde on water related properties of solid wood. *Holzforschung* 64(4): 483-488.

[16] Indonesia, S. N. (2014). *Uji ketahanan kayu terhadap organisme perusak kayu (SNI 7207-2014)*. Jakarta: Badan Standarisasi Nasional.

[17] Karnasudirdja S., Sofyan K., & Kusumodiwiryo. R. (1974). *Pedoman pengujian sifat fisik dan mekanik kayu. Publikasi Khusus No. 20.* Bogor: Lembaga Penelitian Hasil Hutan.

[18] Boer, E. & Ella, A.B. 2001. Plant Resources of South-East Asia No. 18: Plants producing exudates. Bogor, Prosea.

[19] Farag, Y. 2010. *Characterization of different shellac types and development of shellac coated dosage forms* (Doctorate dissertation). Hamburg. Hamburg University.

[20] Hill, C.A.S. & Papadopoulos, A.N., 2001. A review of methods used to determine the size of the cell wall microvoids of wood. *Journal of the Institute of Wood Science* 90(6): 337-345.

[21] Kato, K.L. & Cameron, R.E. 1999. A review of the relationship between thermally-accelerated ageing of paper and hornification. *Cellulose* 6(1): 23-40.

[22] Arinana, Tsunoda K., Herliyana E.N. & Hadi Y.S. 2011. Determination of resistance of woods against subterranean termites by laboratory tests using Indonesian and Japanese standards. Pp 126–131 in *Proceedings of the 8th Conference of the Pacific Rim Termite Research Group.* 28 February–1 March 2011, Bangkok.

IOP Conf. Series: Earth and Environmental Science **359** (2019) 012002 doi:10.1088/1755-1315/359/1/012002

Effect of Clone and Sites for Wood Properties of Fast Grown Teak and Determination of The Best Clone for Next Cultivation in Teak Plantation Industries

R. Damayanti[1*], J Ilic[2], B Ozarska[3], G Pari[4] and P Vinden[5]

[1]Forest Product Research and Development Center, Research, Development and Innovation Agency (FORDIA), Ministry of Environment and Forestry, Bogor, Indonesia
[2]Dept. of Forest and Ecosystem Sciences, The University of Melbourne, Burnley Campus, Richmond, Victoria, Australia
[3]Dept. of Forest and Ecosystem Sciences, The University of Melbourne, Burnley Campus, Richmond, Victoria, Australia
[4]Forest Product Research and Development Center, Research, Development and Innovation Agency (FORDIA), Ministry of Environment and Forestry, Bogor, Indonesia
[5]Dept. of Forest and Ecosystem Sciences, The University of Melbourne, Burnley Campus, Richmond, Victoria, Australia

*email: ratih_turmuzi@yahoo.com

Abstract. Cloning copies all of the characteristics of an individual, not just the optimal characteristics, and thus the clonal plantations have a vital role in increasing forest productivity. Determination of the surviving clone with the best properties in all plantation sites may assure uniformity in wood properties. Determination of the surviving clone with the best properties in all plantation sites may assure uniformity in wood properties. The more uniform the wood properties, the higher price the industries are willing to pay. This study aimed to investigate wood properties variation within and between selected clones at two different plantations of super teak, in order to investigate the effect of genetic or environment on wood properties and to determine the best clone for subsequent cultivation based on the wood properties. Research was conducted on three clones of 5-year old fast grown teak at two locations. The results showed that genetic factor more influenced the uniformity of wood color. Furthermore, site effect was more dominant on wood density, wood hardness, and radial unit shrinkage, while MOE, fibre length, and 'crystallinity' were affected by genetic effect. Effect of clone was not clear on radial and tangential shrinkage, tangential unit shrinkage and MOR. In the case of fast grown teak, Clone B can be recommended for future plantation at both sites.

1. Introduction

Teak (*Tectona grandis* Linn. f) is a popular and highly desirable timber for furniture, flooring, construction building, poles, boat making and appearance grade veneer. Recent data shows that globally, the demand of plantation teak increased 47% by volume and 58% by value for the period of 2010 - 2012 [1]. Mature teak is mostly harvested at the age of approximately 80 years [2]. The rotation age of established plantation forests from India and Indonesia is 50 to 80 years [3]. As a result of high demand, this long rotation has caused the price of teak wood to increase significantly due to a limited supply. Consequently, timber industries that rely on teak as a source of raw material face difficulties with the continuity of stock supply [4].

This situation has prompted the silviculturists to investigate various methods which would allow establishing shorter rotation stands and a faster growth of teak. One of the methods already developed is through vegetative cultivation, such as tissue culture, bud grafting and shoot cutting. As a result, the rotation age can decrease from 50 - 80 years to 20 - 40 years [5]. The new concept of fast growth with high productivity management of teak plantations is widely developed in the world, and it now becomes a general trend [6].

In Indonesia, many types of growing regimes of fast growing teak have been widely cultivated. The combination of breeding technology and intensive silviculture treatment has enabled teak timber producers to harvest the tree at a very young age, for example as early as 5 years. Timber communities call this timber "super teak".

The super teak was planted from several selected mother trees (clone). Clonal plantations have a vital role in increasing forest productivity [7]. Individual selection based on higher growth rate of clones may accede to the use of younger-age trees and consequently reducing the plantation harvesting cycle [8].

Cloning copies all of the characteristics of an individual, not just the optimal characteristics [9]. Furthermore, it is not possible to select only one trait for a specific site if the clone is already proven to survive and provide the desired wood properties. Thus, looking for groups of clones with similar characteristics that can be planted in small areas with specific site conditions is desired [9].

Determination of the surviving clone with the best properties in all plantation sites may assure uniformity in wood properties. The more uniform the wood properties are, the higher price the industries are willing to pay [10]. The first section of this paper reports on the investigation of the wood properties variation within and between selected clones at two different plantations of super teak, in order to determine the effect of genetic or environment on wood properties. The second section of this study aimed to determine the best clone for subsequent cultivation based on the wood properties investigated: wood colour, heartwood proportion, density, shrinkage, unit shrinkage, hardness, bending properties (modulus of elasticity (MOE) and modulus of rupture (MOR)), fibre length, micro fibril angle (MFA), and degree of crystallinity.

2. Materials and methods

2.1. Materials

Six trees from large diameter stems (diameter > 26 cm) of 5- year old super teak plantation forests in Bogor, West Java, Indonesia, and Magetan, East Java in Indonesia were selected and felled (**Error! Reference source not found.**). Trees SA22 and AC23 were from Clone A; both trees were planted in the wet site (Bogor). Furthermore, trees AC18 and 198 were originated from Clone B and were planted at different sites (Tree AC18 in the wet site while tree 198 in the dry site). Trees 12 and 101 were from Clone B; both trees were planted in the dry site (Magetan). Evaluation within clone was carried out at the same plantation sites (Clone A in the wet site and Clone C in the dry site), and at different plantations (Clone B planted at wet and dry sites).

Table 1. Tree selection used in the study.

Tree code	Diameter (cm)	Height* (m)	Clone**	Tree code	Diameter (cm)	Height* (m)	Clone**

Site 1 (West Java)				Site 2 (East Java)			
SA22	30.25	16.0	A	198	29.3	10.5	B
AC18	31.4	18.2	B	12	29.3	12	C
AC23	32.2	13.6	A	101	28.7	9.2	C

*Economic height: height of stem at the top with minimum diameter of 7 cm

2.2. *Methods*

2.2.1. Wood properties measurement and analysis. The wood properties investigated were wood colour, heartwood proportion, density, shrinkage, unit shrinkage, hardness, bending properties (MOE and MOR), fibre length, MFA, and degree of crystallinity.

2.2.1.1. Colour and heartwood measurement. Three, five centimetre thick discs were taken from the bottom, middle, and top of each tree. In addition to colour measurement, specimens were also taken from sawn boards between the bottom and middle sections to observe the colour variation from the tangential faces (along the length of boards).

2.2.1.1.1. Colour measurement. A “CR10” Tristimulus colourimeter (Konika Minolta®) with an 8 mm aperture at the point measurement was used for measuring the colour coefficients of the sapwood, transition-wood and heartwood zones from cross sections of disc samples, and also from the sapwood and heartwood zones of the sawn boards. Three replicate measurements were conducted at different locations from each zone. All colour measurements were made using standard illumination conditions: D65 and 10 degrees observer [11]. Small circular areas were marked on the faces of test specimens to ensure that repeatable measurements were performed at the same location. The colour coefficients L a*b* were measured and used to represent wood colour of each sample. All measurement points were treated as target (absolute measurements).

2.2.1.1.2.Heartwood measurement. To minimize the human error, automatic image processing was used to determine heartwood percentage using ImageJ® freeware routines [12]. A digital camera was employed to capture images from each sanded disc. Total disc area was determined by thresholding black/white (B/W) images of disc. The heartwood area was conveniently further threshold due to grey level differences from the background wood to provide proportions of heartwood from each disc.

2.2.1.2. Density, shrinkage and unit shrinkage. The discs (seven discs from each tree) were used for shrinkage, density, crystallinity, and hardness measurements. Shrinkage samples, measuring 25 mm (radial) x 25 mm (tangential) x 100 mm (longitudinal) were prepared from each disc aiming to preserve true radial and tangential directionality [13]. Specimen’s radial position was also denoted as heartwood (HW), transition-wood (TZ) and sapwood (SW) accordingly. Six replications for each zone (depend on disc/stem diameter) were used. The procedures used for determining densities and shrinkage properties were carried out largely according to Kingston and Risdon [13].

The shrinkage properties in this paper refer to shrinkage and unit shrinkage. Shrinkage occurs when bound water is removed from the cell walls in a condition below the fibre saturation point (FSP) [14]–[16]. The magnitude of wood shrinkage is often used as a guide for anticipated stability in service

[17]. It influences the behavior of the timber during seasoning and affects the target size of sawn green timber [16].

Wood shrinks or swells in three principal directions: longitudinal, radial and tangential. In normal wood, longitudinal shrinkage is one or two orders of magnitude less than transverse shrinkage, and tangential shrinkage is usually 1.5 to 2.5 times higher than radial shrinkage [18].

Some researchers have used the ratio of tangential to radial (T/R) as a descriptor of dimensional stability of wood [19]–[22]. Reportedly the T/R ratio for wood ranges from 1.40 to more than 2, and the wood best suited for uses is one with a low T/R ratio [23].

However, Brown et al. [24] indicate that the T/R ratio might be more or less independent of structure, and that it tends to be related to anisotropy [22]. Kingston and Risdon [13] recommend the use of unit shrinkage, as a good indicator of the potential for the timber to move in service. Kingston and Risdon [13] derived the unit shrinkages from the ratio of the difference in shrinkage between 12% and 5% EMC over the change in moisture content.

The standard procedure for determining moisture content of timber was applied according to AS/NZS 1080.1: 2012 "Timber – Methods of test – Moisture Content" which prescribes that the end point condition is reached when the weight change is less than 0.2% of the previous determination. The radial and tangential dimensions were measured using digital calliper (Mitutoyo®) with 0.001 mm accuracy. The tangential and radial measurements were taken at three equidistant points along each sample in the middle of the faces. Three measurements for each dimension were averaged for each test specimen. The lengths of the test specimens were measured with digital vernier calliper (Mitutoyo®) with 0.01 mm accuracy.

Test specimen weights were made to an accuracy of 0.001 g. The densities measured were basic density (oven dry mass / green volume), air dry density (mass at 12% MC / volume at 12% MC before and after steam reconditioning) and oven dry density (mass at 0% MC / volume at 0% MC) [13]. Shrinkage values from green to 12% MC, 5% MC and 0% were expressed as a percentage of green dimensions. Unit shrinkages were calculated by dividing the difference in the shrinkage of the specimens at the nominal 12% and 5% EMC by the difference in the equilibrated moisture content for each individual specimen.

2.2.1.3. Hardness. The specimens used were the same as the density and shrinkage specimens (2.2.1.2), but, the opposite side of each disc was used for hardness measurement. The hardness was determined using an Instron® universal strength testing machine according to the Australian Standards for mechanically testing small clear specimens of timber [25]. Wood specimens were tested in the radial (R hardness), tangential (T hardness) and longitudinal (end hardness) grain directions. The reason for testing all wood faces was to ensure a more representative value of each specimen [26], and to enable radial and tangential hardness to be tabulated as side hardness for ready comparison with data for other species. Before and after testing, the weight and dimensions of each specimen were measured for the determination of wood density and moisture content.

2.2.1.4. Bending properties. Specimens for bending tests were taken from bottom (Log 1), middle (Log 2), and top part (Log 3) of the trees, and grouped into heartwood, transition-wood and sapwood zones. Three replicates were used for each axial and radial position. 20 x 20 x 300 mm specimens were prepared for bending tests employing an Instron® universal strength testing machine according to the Australian Standards procedure for mechanically testing small clear specimens of timber [25].

The loading rate was 1.0 mm per-minute and the specimens were loaded on the radial face. Before and after testing, the weight and dimensions of each specimen were measured for determination of wood density and moisture content.

2.2.1.5. Fibre length. Three, five-centimeter thick discs were taken from bottom, middle and at the top (stem diameter minimum 7 cm) of each tree, and separated into three zones along the radial direction: heartwood, transition-wood zone and sapwood. Specimens were chipped before maceration. The maceration process was carried out according to the Franklin method described in Tesoro [27]. The samples were heated slowly at 40 – 60 °C in the mixture of Acetic Acid and Hydrogen Peroxide (H_2O_2) at a ratio of 1 : 1 in a boiling water bath. The heating process took about 12 hours to produce milky white to almost transparent whips, these being adequately macerated for a satisfactory separation of cells. The separated cells were washed with water to remove residual acid and H_2O_2, and stained with Safranin.

The fibres were randomized by using a method described by Assoc. Prof Jugo Ilic. A plastic petri dish (the diameter slightly larger than the length of a slide) was perforated with one mm holes 1 cm apart all over the bottom, was loaded with a clean glass slide, and placed in a container filled with water up to the bottom of the petri dish. Slurry of fibres then were added into petri dish and topped up with water, and then stirred gently to randomize the fibres. The water was released gradually from the larger container until the water level was lower than the bottom of the glass slide. Slides with randomized exposed fibres were covered with glycerol or permanent mount in a few cases for further reference. The length of 25-fibres was measured for each zone: heartwood, transition-wood and sapwood.

2.2.1.6. Microfibril Angle (MFA) and Degree of Crsytallinity. Specimens for the crystallinity measurement were end-matched with the density and shrinkage specimens. The samples were 50 mm (tangential direction) x 15 mm (longitudinal direction) x 30 - 40 μm (radial direction). The crystallinity and its quantifiers were measured using X-Ray Diffraction (XRD) (Shimadzu®). The X-ray beam was powered with a 40kV, 30mA source and scans made in the range from 0 - 40 degrees at a scan speed of 2 degrees per minute.

The degree of crystallinity (DC) was calculated as the ratio between the crystalline and total non-crystalline and amorphous regions. The dimension of crystallites (CW-width and CL-length) was determined using the Scherrer formula [28] for diffraction intensities (002) and (040), usually in the angular range from 20 to 24 degrees and from 32 to 37 degrees, respectively [29]. Cellulose lattice spacing (d) was calculated using Bragg Equation [29]. The number of crystallite planes for each of the diffraction intensities was calculated by dividing the crystallite dimension with the crystallite lattice spacing (d).

Micro fibril angle (MFA) was determined from the (002) reflection plane, and calculated by using Cave's method. The T parameter was calculated according to Stuart and Evans [30] by manually drawing tangents to the sides of plots of the diffraction arcs.

2.2.2. Wood properties variation within clone. Variation within a clone was assessed by comparing the wood properties of different trees originated from the same clone.

2.2.3. Wood properties between clones. Variation between clones was assessed by comparing different clones planted at the same site: Clone A and Clone B at the wet site, then Clone B and Clone C at the dry site. Both variations within clone and between clones were tested using independent t-test if the data were normally distributed and Mann-Whitney U Test if the data were not normally distributed. The distribution of wood properties data was tested according to Saphiro-Wilk's W test (Statistica®).

2.2.4. Determination of the best clone. For the determination of the best clone, wood properties of each clone were averaged and tabulated. Assessments were then conducted based on the best wood properties (compared to mature teak). Clone with the closest wood properties compared to mature teak was each allocated one point. The points for each clone then were added, and the clone with the highest score was selected as the preferred clone.

3. Results and discussion

3.1. Variation of wood properties within and between clones

3.1.1. Wood colour and heartwood proportions. In this study, super teak trees from the same clone planted on the same site exhibited similar colour coefficients, except for the yellowness index for trees at the wet site. Furthermore, if the same clone was planted on different sites, Clone B exhibited similar a* and b* coefficients. **Error! Reference source not found.**2 shows the variation of the colour index within clone planted at same and different sites. The result is also in good agreement with the report from Gierlinger [31] that the wood colour of trees from the same provenance grown on different sites is similar. And this is consistent with the result of a study by Moya and Marin [9], showing that heartwood colour is significantly dependent on clone type.

Table 2. Colour variation within clones of super teak

Variables	Clone A (Wet Site)			Clone B (Wet and Dry Sites)			Clone C (Dry Site)		
	SA22	AC23	P Value	AC18	198	P Value	12	101	P Value
L	61.2	63.4	NS	61.5	56.9	0.02*	62.6	61.1	NS
a (+)	9.2	9.7	NS	10.5	11.0	NS	10.1	11.4	NS
b (+)	19.3	21.5	0.00**	22.2	21.8	NS	24.0	25.1	NS

*=Significant at α=<0.05; **= Significant at α=<0.01 according to independent t-test or Mann-Whitney U test; SA22, AC23, AC18, 198, 12 and 101 are tree codes; SA22 and AC 23 from Clone A; AC18 and 198 from Clone B; and tree no 12 and 101 are from Clone C

Table 3. Color variation between clones of super teak.

Variables	Wet Site			Dry Site		
	Clone A	Clone B	P Value	Clone B	Clone C	P Value
L	62.4	61.5	NS	56.9	61.7	0.02*
a (+)	9.5	10.5	NS	11.0	10.9	NS
b (+)	20.5	22.2	0.00*	21.8	24.6	0.00*

*=Significant at α=<0.05; **= Significant at α=<0.01 according to independent t-test or Mann-Whitney U test

This study revealed that trees from different clones planted at the wet site resulted in different b* and exhibited different L and b* values when planted at the dry site. Table 33 shows the variation of the colour index between different clones planted at the same sites. These results support the finding by Gierlinger [31] who observed that Larch (*Larix* spp.) trees from different provenances planted on the same site showed different colour characteristics.

The colour characteristics especially lightness and yellowness of wood from different clones were evident even if the clones were planted on the same site. This study showed that the effect of clone was greater on wood colour than the effect of site. This result is in line with Moya and Marin [9]. Thus, it is suggested that if specific colour preference is needed of plantation trees, clone selection is important.

Unfortunately, the analysis of variance of heartwood proportion from clones could not be applied as there were only three to six discs from each clone. this is in line with the colour properties by Moya and Marin [9] reported that the heartwood proportion from young teak was more influenced by clone than site.

3.1.2. Density, shrinkage, unit shrinkage and mechanical properties

Table 44 shows variations within clone of oven dry density, shrinkage, unit shrinkage and mechanical properties of super teak. Trees from the same Clone A and Clone B planted at a same site generally showed similar wood density (see density of shrinkage and mechanical specimens in Table 4) when planted at the wet site (Clone A), but different wood densities when planted at the dry site (Clone B). However, when a same clone planted at different wet and dry sites (Clone B), the density was consistently and significantly different. According to Saranpää [32], the basic density was under a strong genetic control. However in this study, site effects were more dominant than genetic effects, and this result was in good agreement with Zobel and Buijtenen [33] who reported that the inheritance of wood density in teak was rather weak. Indira and Bhat [34] also reported that site had a great influence on wood density.

Different patterns are observed for the shrinkage and unit shrinkage variations within clone. In general, trees from the same clone planted at the same and different sites exhibited similar shrinkages. Thus, genetic effect seemed to be more dominant for shrinkage than the site effect. Unfortunately except for longitudinal shrinkage in softwoods, published information is difficult to find on the effect of genetics on transverse wood shrinkage.

In unit shrinkage, trees from the same clone planted at the wet site (Clone A) showed similar radial unit shrinkage but different tangential unit shrinkage. Furthermore, when planted at dry site (Clone C) and at different sites (Clone B), different results in radial unit shrinkage but similar tangential unit shrinkage were obtained. For radial unit shrinkage, site effect was more dominant, but in tangential unit shrinkage, genetic effect was more dominant.

Hardness showed slightly different patterns than density. At the wet site, trees from the same clone exhibited little difference in wood hardness. However, when planted on the dry site, even trees from the same clone showed consistently different wood hardness. Furthermore, when trees from the same clone were planted at different sites, they showed similar side hardness but different end-grain hardness. Only end-grain hardness followed the pattern in the density. Hence in this study, no clear variation within clone on side-hardness was observed.

Table 4. Within clones variation of density, shrinkage and mechanical properties of super teak.

Variables	Clone A (Wet Site)			Clone B (Wet and Dry Sites)			Clone C (Dry Site)		
	SA22	AC23	P Value	AC18	198	P Value	12	101	P Value
Oven dry density (kg m^{-3})	483	479	NS	470	514	0.00*	476	514	0.00*
Shrinkage from green to 0% MC (%)									
Radial	2.8	3.0	NS	2.9	2.9	NS	2.9	3.2	NS
Tangential	4.4	5.0	0.03*	5.1	5.1	NS	5.3	5.4	NS
Unit Shrinkage									
Radial	0.13	0.14	NS	0.13	0.15	0.00*	0.15	0.17	0.02*
Tangential	0.19	0.22	0.05*	0.21	0.22	NS	0.21	0.23	NS
Hardness (N)									
Air dry density (kg m^{-3})	535	518	NS	502	541	0.00**	500	540	0.00**
Radial	3132	2922	NS	2963	3314	NS	2785	3298	0.00**
Tangential	3529	3443	NS	3415	3554	NS	2843	3350	0.00**
End-grain	4323	4024	NS	3839	4440	0.00**	4137	4704	0.00**
MOE & MOR									
Air dry density (kg m^{-3})	574	500	0.00*	518	578	0.00*	537	558	NS
MOE (GPa)	8.5	7.9	NS	7.4	7.8	NS	7.4	9.0	NS
MOR (MPa)	64.6	47.3	0.01*	47.9	49.3	NS	57.6	60.9	0.00*

*=Significant at α=<0.05; **= Significant at α=<0.01 according to independent t-test or Mann-Whitney U test; SA22, AC23, AC18, 198, 12 and 101 are tree codes; SA22 and AC 23 from Clone A; AC18 and 198 from Clone B; and tree no 12 and 101 are from Clone C

The same pattern was observed for the variation of MOE within clones. Trees from the same clone planted at same and different sites exhibited similar MOE values, thus, clone effect was more dominant than the site effect. For MOR, trees from the same clone planted at the same site exhibited different values, but when planted at different sites showed similar values. Thus, it is not clear whether genetic effect or site effect had greater effect on MOR.

To clarify the effect of clone or site on the wood properties, variations between different clones planted at the same site were then studied. The results are shown in

Table 5. In general, trees from different clones planted at the same site exhibited similar wood density. This result confirms the above finding that the site effect was more dominant than the genetic effect for wood density.

Table 5. Variation of density, shrinkage, and mechanical properties between clones.

Variables	Wet Site			Dry Site		
	Clone A	Clone B	P Value	Clone B	Clone C	P Value
Oven dry density	515	503	NS	514	497	0.05*
Shrinkage from green to 0% MC						
Radial	2.9	2.9	NS	2.9	3.1	NS
Tangential	4.7	5.1	NS	5.1	5.3	NS
Unit Shrinkage						
Radial	0.13	0.13	NS	0.15	0.16	NS
Tangential	0.20	0.21	NS	0.22	0.23	NS
Hardness (N)						
Air dry density (kg m^{-3})	525	502	0.00*	541	518	0.01*
Radial	3009	2963	NS	3314	3018	NS
Tangential	3478	3415	NS	3554	3073	0.00*
End-grain	4148	3839	NS	4440	4395	NS
MOE & MOR	540	518	NS	578	549	0.01*
Air dry density (kg m^{-3})	8.2	7.4	NS	7.8	8.3	NS
MOE (GPa)	56.6	47.9	NS	49.3	59.5	0.03*
MOR (MPa)	540	518	NS	578	549	0.01*

*=Significant at α=<0.05according to independent t-test or Mann-Whitney U test

For the shrinkage properties (shrinkage and unit shrinkage), the results in

Table 55 show that trees from different clones planted at the same site exhibited consistently similar shrinkage and unit shrinkage values. The previous results in Table 44 shows that clone effect is more dominant than site effect for radial and tangential wood shrinkage and the tangential unit shrinkage, but site effect is more dominant on radial unit shrinkage. The results in Table 5 are not as expected because it appeared that site effect is consistently more dominant on shrinkage properties. Consequently, it is not clear which factor had more dominant effect on shrinkage and tangential unit shrinkage.

Wood hardness of trees from different clones planted at same site exhibited generally similar hardness values. From the above result no clear pattern of variation within clone is observed, and it can be concluded that in line with wood density, the site effect is greater on wood hardness than the genetic effect.

Trees from different clones planted at the wet site showed similar MOE, but when planted at the dry site, the trees exhibited different MOE values. The previous conclusion that MOE was more influenced by genetic effect than by site effect is confirmed by the result from the dry site. Similar variation of MOE is observed with MOR. The above result shows that neither the effect of clone nor site is observed for MOR. This can be seen from

Table 55.

3.1.3. Fibre length. The results presented in Table 6 show that in general, the same clone planted at the same or different site exhibited similar fibre length. Considering fibre length effect due to genetics seemed to be greater than the effect of site is indicated in Table 77; different clones planted at same site exhibit different fibre length values.

Table 6. Within clones variation of fibre length of super teak.

Cell Length (µm)	Clone A (Wet Site)			Clone B (Wet and Dry Sites)			Clone C (Dry Site)		
	SA22	AC23	P Value	AC18	198	P Value	12	101	P Value
Fibre	1391	1303	0.00**	1320	1317	NS	1314	1315	NS

**= Significant at α=<0.01 according to independent t-test or Mann-Whitney U test; SA22, AC23, AC18, 198, 12 and 101 are tree codes; SA22 and AC 23 from Clone A; AC18 and 198 from Clone B; and tree no 12 and 101 are from Clone C

Table 7. Between clones variation of fibre and vessel length of super teak.

Cell Length (µm)	Wet Site			Dry Site		
	Clone A	Clone B	P Value	Clone B	Clone C	P Value
Fibre	1347	1320	0.02*	1317	1314	NS

*=Significant at α=<0.05 according to independent t-test or Mann-Whitney U test

3.1.4. Ultrastructure 'crystallinity'. In general, trees from the same clone planted at the same or different sites exhibited similar 'crystallinity', as indicated in Table 8. The result shows that clone effect is more dominant than factor site. Furthermore, when trees from different clone planted at same site, it also showed similar 'crystallinity' characteristics (Table 9).

Table 8. Variation of wood 'crystallinity' within clone.

Variables	Clone A (Wet Site)			Clone B (Wet and Dry Sites)			Clone C (Dry Site)		
	SA22	AC23	P Value	AC18	198	P Value	12	101	P Value
MFA (°)	18.2	15.8	0.02*	15.5	15.0	NS	14.0	16.3	NS
DC (%)	37.2	46.1	0.00**	42.9	45.7	NS	42.5	45.3	NS
Crystallite Width (nm)	3.7	3.6	NS	3.5	3.6	NS	3.5	3.4	NS
d002	0.39	0.39	NS	0.39	0.39	NS	0.39	0.39	NS
N002	9.3	9.2	NS	9.0	9.1	NS	8.9	8.7	NS
Crystallite Length (nm)	27.3	24.8	NS	29.1	28.1	NS	32.1	26.8	NS
d040	0.25	0.25	NS	0.25	0.25	NS	0.24	0.26	NS
N040	109.4	99.5	NS	117.1	111.9	NS	132.7	104.1	NS

*=Significant at α=<0.05; **= Significant at α=<0.01 according to independent t-test or Mann-Whitney U test; SA22, AC23, AC18, 198, 12 and 101 are tree codes; SA22 and AC 23 from Clone A; AC18 and 198 from Clone B; and tree no 12 and 101 are from Clone C; d = crystallite lattice spacing; N = number of crystallite planes

Table 9. Variation of wood 'crystallinity' between clones.

Variables	Wet Site			Dry Site		
	Clone A	Clone B	P Value	Clone B	Clone C	P Value
MFA (°)	17.1	15.5	0.05*	15.0	15.3	NS
DC (%)	41.5	42.9	NS	45.7	44.2	NS
Crystallite Width (nm)	3.6	3.5	NS	3.6	3.4	NS
d002	0.39	0.39	NS	0.39	0.39	NS
N002	9.3	9.0	NS	9.1	8.8	NS
Crystallite Length (nm)	26.1	29.1	NS	28.1	28.8	NS
d040	0.25	0.25	NS	0.25	0.25	NS
N040	104.6	117.1	NS	111.9	115.0	NS

Note: P values were obtained by using independent t-test or Mann-Whitney U test; SA22 and AC 23 from Clone A; AC18 and 198 from Clone B; and tree no 12 and 101 are from Clone C; d = crystallite lattice spacing; N = number of crystallite planes

3.2. Determination of the"best" clone based on the wood properties

Table 10 presents the results of the assessment of the variation of wood properties examined to determine within clone effects of trees planted at Bogor and Magetan. The wood properties were assessed on the basis of their similarity of super teak to that of approximately 80-year old mature teak as described in Indonesian Wood Atlas [35] and other relevant literature.

Table 10. Summary of the average wood properties of super teak for scoring for best clone based on the similarity with mature teak

Variables	Clone A (Wet Site)			Clone B (Wet and Dry Sites)			Clone C (Dry Site)			Mature Teak
	Mean	STD	COV	Mean	STD	COV	Mean	STD	COV	
Height	15*	2	11	14	5	38	11	2	19	-
Diameter	31*	1	4	30	1	5	29	0.4	1	-
Colour [36]										
L	62*	9	14	59	10	16	62*	6	9	62.6
a*	9	3	30	11*	3	31	11*	3	26	6.38
b*	21	2	11	22	2	10	25*	2	9	25.8
Heartwood %	18	10	56	30*	6	20	20	15	75	84
Density and shrinkage										
Shrinkage from green to 0% MC [3]										
Radial	2.9*	0.4	13	2.9*	0.4	14	3.1	0.5	16	2.5-3.0
Tangential	4.7*	0.9	19	5.1	0.8	16	5.3	0.9	18	3.4-5.8
Unit Shrinkage (no reference was found; the best unit shrinkage was selected as the lowest value)										
Radial	0.13*	0.03	19	0.15	0.03	17	0.16	0.03	18	-
Tangential	0.20*	0.04	22	0.22	0.04	16	0.23	0.05	21	-
Density (kg m^{-3}) [3]										
Air dry (corrected)	515	21	4	532*	39	7	528	36	7	480-850
Oven dry	481	20	4	500*	37	7	497	35	7	-
Hardness (N) [35]										
Side Hardness	3244	678	21	3278*	531	16	3046	417	14	4197
End Hardness	4148	588	14	4124	632	15	4395*	563	13	4059
Stiffness and bending [3]										
MOE (GPa)	8	2	22	8	1	20	8	2	20	8.6-13.4
MOR (MPa)	57	21	37	49	19	38	60*	15	25	85-148
Cell Dimensions (μm) [35]										
Fibre length	1347*	143	11	1319	119	9	1314	111	8	1316
"Crystallinity" [36]										
MFA (°)	17	3	15	15*	2	16	15*	4	24	14-17
DC (%)	41	8	19	44*	7	15	44*	7	17	-
Total score	8			8			7			

*indicates a score of one when the wood properties of a clone was closest to mature teak's property (the best clone was chosen with the highest score based on wood properties)

The results showed that Clone A planted at the wet site exhibits eight points when the assessments were combined between the assessment on the wood properties and the tree size. Consequently, the score for Clone A is six when wood properties only were included.

The score of Clone C is seven; here the score of seven is based on the assessments on wood properties only because the tree size (height and diameter) of trees when planted at the dry site were lower than those Clone A because trees from Clone C were all planted under less favourable conditions than trees from Clone A. Hence, if the assessment is based on wood properties only, Clone C exhibits 'better' wood properties than Clone A. However, the score attributed to Clone C may have been affected by site effect because some wood properties such as hardness is more influenced by site effect (section 3.3.1 Variation of wood properties within and between clones). Accordingly, the scores seven obtained in Clone C possibly are achieved not only because of clone effect itself but also because they were planted at drier environment, for example wood hardness was 18% higher at the dry site compared to the wet site.

Clone B which attained a score of eight appeared to have the highest wood properties compared to the other clones. The tree sizes are also comparable to the trees from Clone A planted at the wet site. From the result, it has been proven that even though planted at different sites, Clone B still exhibits comparable wood properties. Based on the score and validity of the evaluation, it would appear that Clone B can be suggested for planting at both sites for next cultivation.

4. Conclusions

A uniform wood colour was observed in trees from the same clone irrespective of growing site. Site effect is more dominant on wood density, wood hardness and radial unit shrinkage, while MOE, fibre length, and 'crystallinity' were affected by genetic effect. Effect of clone type was not clear on radial and tangential shrinkage, tangential unit shrinkage and MOR. For the best clone, Clone B can be recommended for future replanting at both sites.

Acknowledgement

Thanks to ACIAR (Australian Centre for International Agricultural Research) for giving funding of this work. Thank you are also presented to the technicians: Tutiana, Romi, and Ida who have helped in the data collection. Bpk. Hariyono Suroso (RIP) and Bpk. Muhammad Rafik (Unit Bagi Hasil KPWN Ministry of Forestry, Indonesia) are gratefully acknowledged for helping in providing the experimental materials – Jati Utama Nasional. All authors are equally contributed to the paper writing.

References

[1] S. Midgley, R. T. Somaiya, P. R. Stevens, A. Brown, N. D. Kien, and R. Laity, "Planted teak: global production and markets, with reference to Solomon Islands," Australian Centre for International Agricultural Research (ACIAR), Canberra, 2015.

[2] M. Trockenbrodt and J. Jouse, "Wood properties and utilisation potential of plantation teak (Tectona grandis) in Malaysa-a critical review," *J. Trop. For. Prod.*, vol. 5, p. 13, 1999.

[3] I. Soerianegara and R. Lemmens, *Plant Resources of South-East Asia 5 (1). Timber trees: Major commercial species*. Leiden: Backhuys Publisher, 1994.

[4] Krisdianto and G. Sumarni, "Perbandingan persentase volume teras kayu jati cepat tumbuh dan konvensional umur 7 tahun asal Penajan, Kalimantan Timur ((Heartwood Portion in Logs of 7 years old Fast Growing and Conventional Teak taken from Penajam, East Kalimantan)," *J. Penelit. Has. Hutan*, vol. 24, no. 5, pp. 385–394, 2006.

[5] A. D. Yunianti, "Nanostructure Characteristics of Cell Wall and their Relationship to Wood Quality (Case study in 7-year old cloned teak)," Bogor Agricultural University, Bogor, 2012.

[6] I. Bermejo, I. Cañellas, and A. S. Miguel, "Growth and yield models for teak plantations in Costa Rica," *For. Ecol. Manage.*, vol. 189, no. 1–3, p. 19, 2004.

[7] K. Palanisamy, K. Gireesan, V. Nagarajan, and M. Hegde, "Selection and clonal multiplication of superior trees of teak (Tectona grandis) and preliminary evaluation of clones," *J. Trop. For. Sci.*, vol. 21, no. 2, p. 7, 2009.

[8] K. M. Bhat, P. B. Priya, and P. Rugmini, "Characterisation of juvenile wood in teak," *Wood Sci. Technol.*, vol. 34, p. 16, 2001.

[9] R. Moya and J. D. Marin, "Grouping of Tectona grandis (L.f.) clones using wood color and stiffness," *New For.*, vol. 42, no. 3, p. 17, 2011.

[10] R. E. Vetter, V. R. Coradin, and E. C. Martino, "Wood colour - a comparison between determination methods," *IAWA Bull.*, vol. 11, no. 4, p. 11, 1990.

[11] HunterLab, "Measuring Color using Hunter L, a, b versus CIE 1976 L*a*b*," vol. AN 1005.00. HunterLab, Reston, USA, 2012.

[12] M. Mekhtiev and G. Torgovnikov, "Method of check analysis of microwave-modified wood," *Wood Sci. Technol.*, vol. 38, p. 13, 2004.

[13] R. S. T. Kingston, C. J. E. Risdon, and C. S. and I. Organization, "Shrinkage and Density of Australian and other South-West Pacific Wood," Division of Forest Product, Australia, 1961.

[14] A. J. Stamm, *Wood and Cellulose Science*. New York: New York: Ronald Press Co., 1964.

[15] J. C. F. Walker and R. Woollons, "Cell wall organisation and the properties of xylem - a speculative review," in *IUFRO/IAWA International Workshop on the significance of microfibril to wood quality*, 1998, pp. 13–26.

[16] J. L. Bowyer, R. Shmulsky, and J. G. Haygreen, *Forest Products and Wood Science An Introduction*, 4th ed. Iowa: Iowa State Press (Blackwell Publishing Company), 2003.

[17] D. J. Cown, D. L. McConchie, and G. D. Young, *Radiata Pine Wood Properties Survey*. Rotorua New Zealand: Forest Research Institute Ministry of Forestry, 1991.

[18] J. C. F. Walker, B. G. Butterfield, T. A. G. Langrish, J. M. Harris, and J. M. Uprichard, *Primary Wood Processing: Principles and Practice*. London: Chapman & Hall, 1993.

[19] E. Basri, "Bagan pengeringan dasar 12 jenis kayu dari Indonesia (Basic drying schedules of 12 Indonesian wood species)," *J. Penelit. Has. Hutan*, vol. 26, no. 2, 2008.

[20] E. Basri and S. Rulliaty, "Pengaruh sifat fisik dan anatomi terhadap sifat pengeringan enam jenis kayu (The effect of physical and anatomical properties on drying properties of six wood species)," *J. Penelit. Has. Hutan*, vol. 26, no. 3, p. 17, 2008.

[21] E. Basri, T. A. Prayitno, and G. Pari, "Pengaruh umur pohon terhadap sifat dasar dan kualitas pengeringan kayu waru gunung (Hibiscus macrophyllus Roxb.) (Effect of tree age on basic properties and drying quality of waru gunung (Hibiscus macrophyllus Roxb.))," *J. Penelit. Has. Hutan*, vol. 30, no. 4, p. 11, 2012.

[22] M. Anish, E. Anoop, R. Vishnu, B. Sreejith, and C. Jijeesh, "Effect of growth rate on wood quality of teak (Tectona grandis L. f.): a comparative study of teak grown under differing site

quality conditions," *J. Indian Acad. Wood Sci.*, vol. 12, no. 1, p. 8, 2015.

[23] A. J. Panshin, C. de Zeeuw, and H. P. Brown, *Textbook of Wood Technology. Volume I: Structure, identification, uses, and properties of the commercial woods of the United States*. New York: McGraw-Hill Book Company, 1964.

[24] H. P. Brown, A. J. Panshin, and C. C. Forsaith, *Textbook of Wood Technology Volume II: The Physical, Mechanical, and Chemical Properties of the Commercial Woods of the United States*, vol. II. McGraw-Hill Book Company, 1952.

[25] J. J. Mack, *Australian Methods for Mechanically Testing Small Clear Specimens of Timber*, no. 31. Australia: Commonwealth Scientific and Industrial Research Organization (CSIRO), 1979.

[26] D. Green, M. Begel, and W. Nelson, "Janka hardness using non standard specimens," Forest Product Laboratory, Wisconsin, 2006.

[27] F. Tesoro, "Wood Structure and Quality: Bases for Improved Utilization of Timbers," in *The Second Pacific Regional Wood Anatomy Conferences*, 1989.

[28] M. Peura, M. Muller, U. Vainio, M. P. Saren, P. Saranpaa, and R. Serimaa, "X-ray microdiffraction reveals the orientation of cellulose microfibrils and the size of cellulose crystallites in single Norway spruce tracheids," *Trees*, vol. 22, p. 13, 2008.

[29] K. Abe and H. Yamamoto, "Mechanical interaction between cellulose microfibril and matrix substance in wood cell wall determined by X-ray diffraction," *J Wood Sci*, vol. 51, no. 4, p. 5, 2005.

[30] S. A. Stuart and R. Evans, "X-ray diffraction estimation of the microfibril angle variation in eucalypt increment cores," The CRC for Hardwood Fibre & Paper Science, Australia, 1994.

[31] N. Gierlinger *et al.*, "Colour of larch heartwood and relationships to extractives and brown-rot decay resistance," *Trees*, vol. 18, no. 1, p. 7, 2004.

[32] P. Saranpää, J. A. Roberts, and P. N. R. Usherwood, "Wood density and growth," in *Wood Quality and Its Biological Basis*, no. Bilogical Science Series, J. R. Barnett and G. Jeromidis, Eds. Garsington, UK and Carlton, Australia: Blacwell Publishing Ltd, 2003, p. 226.

[33] B. J. Zobel and J. P. Buijtenen, *Wood Variation: Its Causes and Control*. Berlin, Heidelberg : Springer Berlin Heidelberg, 1989., 1989.

[34] E. P. Indira and K. M. Bhat, "Effects of site and place of origin on wood density of teak (Tectona Grandis) clones," *J. Trop. For. Sci. 6/1/1998, Vol. 10, Issue 4, p. 537-541*, vol. 10, no. 4, p. 5, 1998.

[35] A. Martawijaya, I. Kartasujana, K. Kadir, and S. A. Prawira, *Atlas Kayu Indonesia Jilid I (Indonesian Wood Atlas)*, vol. I. Bogor: Forest Products Research and Development Centre, Indonesian Ministry of Forestry, 2005.

[36] I. Wahyudi, "Study on growth and wood qualities of tropical plantation species," Nagoya University, Japan, 2000.

Growth performance of *Acacia mangium* provenances in Parung Panjang, Bogor and its correlation with physical and mechanical wood properties

Nurhasybi[*] and DJ Sudrajat[**]

Forest Tree Seed Technology Research and Development Institute,
Jl. Pakuan Ciheuleut PO BOX. 105, Bogor, Indonesia

email: [*]d_hasybi@yahoo.com, [**]djsudrajatbtp@yahoo.com

Abstract. Wood quality and growth are two important parameters that need to be analyzed for the utilization and selection of superior genotypes. The objective of this research was to investigate the ten provenances growth characteristics and their physical and mechanical properties of *Acacia mangium* at Parung Panjang, Bogor. The provenance trial was arranged in three blocks of a randomized completely blocks design. In every provenance, 18 trees were sampled resulting in a total of 180 trees in each block representing each provenance measured for their total height, clear bole height, stem diameter at breast height, stem form, branching system, and crown form. Three individual trees were selected randomly from each provenance for wood properties traits, i.e. moisture content, density, modulus of elasticity (MOE) and modulus of rapture (MOR). Significant differences among provenances were detected for stem diameter, clear bole height, branching system and crown form, while total height and stem form revealed not significantly different. Most of growth parameters was not significantly correlated with physical and mechanical wood traits, except for MOR, it revealed significant correlation with the total tree height, stem form, and branching system. Generally, some growth characteristics were not a good indicator for selecting the best provenance in wood properties. The best growth performance was revealed by Kiriwo/Serisa WP, Bimadebun Village, and Balimo District, but from wood quality, the superior provenance was revealed by Bimadebun Village with the lower moisture content, the highest wood density, the highest MOE and MOR. The Bimadebun Village provenance is potential to be developed as provenance seed stand, especialy for solid (contruction) wood.

1. Introduction

Acacia mangium, although turned out to be prone to a number of diseases [1,2], but it is still planted for pulp and paper production in most of forest plantation areas in Indonesia. Total area of plantation forest of this species in the country has reached 1.78 million ha [3]. The species is still promising, partly because of its far superior growth, wide site suitability, multiple uses and also it was easy to plant large areas as planting material was not difficult to obtain [4]. *A. mangium* wood makes attractive furniture and cabinets, molding, door and window components [5]. The wood is also suitable for light structural works, agricultural implements, boxes, crates [6], and decorative veneers [4]. At the present, the species have been planted in community forest especially in Java, for fulfilling the necessary of construction materials.

Acacia mangium has a fragmented natural distribution stretching from the Moluccas islands in Indonesia to Western Province of Papua New Guinea and northeastern Queensland in Australia. Many provenances that highly adapted to their natural habitats have been identified and studied, and they have shown variation among them in all respects [6]. International provenance trials were established during the 1980s and some trials revealed highly significant provenance differences in growth trait among experimental sites and provenance regions were observed [7]. For example, growth was generally faster at near-equatorial trial sites with mean annual height increment around 3-4 m, and slower at sites further from the equator. Papua New Guinea provenances were consistently the best performers, closely followed by the Claudie River provenance from north Queensland. The slowest growing provenances were from the Maluku province of Indonesia and southern parts of the distribution in Queensland. Other studies have given similar results that the PNG provenances had better growth performance than the Queensland provenances [8-10]. In wood characteristics, provenances were significantly affected to wood physical characteristics of *A. mangium*, but they were not significant on mechanical wood characteristics [11].

One of the main factors driving wood quality property difference is growth rate [12]. It is considered to influence wood density in a way that varies among the major tree groups (softwoods, ring-porous and diffuse-porous hardwoods). In the two groups of hardwoods, wood is affected differently by growth rate. In ring-porous hardwoods most studies indicated that high wood density is associated with fast growth [12-14]. On the other study on *Eucalyptus camaldulensis*, there were no significant differences in basic, oven-dry and green density values between fast and slow growth [15]. Growth rate differences have little effect on wood property of hardwoods [16-17], however, a quantitative assessment of the relationship between growth and wood quality is needed in the effective forest management [18]. Wood quality and growth are two important parameters that need to be analyzed for the utilization and selection of superior genotypes [19].

The objective of this research was to investigate the provenances growth characteristics and their physical and mechanical wood properties of *A. mangium*. A better understanding on the relationship between the provenances and wood quality will be helpful in making silvicultural strategies to produce high quality wood by the selecting the best provenance for establishing forest plantations.

2. Materials and methods

2.1. Materials

Growth and wood quality evaluation were conducted on 10 years of *A. mangium* provenance trial at Parung Panjang Forest Research Station (lat. 106° 6‘, long. 06° 20‘) Bogor, West Java. Ten provenances were evaluated for growth and wood properties. Table 1 presented the list of the 10 *A. mangium* provenances from Australia Tree Seed Centre, CSIRO Division of Forestry, collected from Queensland (Australia) and Papua New Guinea (PNG). The altitude of the planting site at Parung Panjang Forest Research Station is 51 m above sea level. Soil type is podzolic haplic with pH 4.8. The soil fertility, organic matter, nitrogen, phosphor, and potassium contents are low. The climate falls into type A according to Schmidt and Ferguson classification, with an annual rainfall between 2000 and 2500 mm (dry season 6-8 months/year) [20]. The original vegetations on the site were dominated by *Imperata cylindrica* grassland and *Schima wallichi* coppices.

Table 1. List of *A. mangium* provenances in the provenances trial at Parung Panjang, Bogor

Provenance code	Origin of provenances	Latitude (S)	Longitude (E)	Altitude (m asl)
16937	Balimo District, PNG	8°17‘	143°06’	30
16938	Kini WP, PNG	8°05‘	142°58’	12
17866	Lake Murray, PNG	6° 51‘	141°29’	55
17946	Claudia River, Queensland	12°48‘	143°18’	20
18057	Kuru, PNG	8°52‘	143°05’	30
18204	Bimadebun Village, PNG	8°38‘	142°03’	40
18206	Arufi Village, PNG	8°43‘	141°55’	25
18209	Kiriwo/Serisa WP, PNG	8°25‘	141°30’	45
18216	Wipin District-PNG	8°47’	142°52’	45
18207	Mata Area WP-PNG	8 °41’	141 °51’	20

Source: Australia Tree Seed Centre, CSIRO Division of Forestry (file no. DA1/145)

2.2. Experimental Design and Growth Measurement

The *A. mangium* provenance trial approximately covers 3 hectares was planted in a randomized complete block design with 3 blocks and the spacing was 3 m x 3 m. In each block, provenance was represented by one plot of 40 trees (5 trees x 8 trees). In every provenance, 18 trees in the middle of plot were sampled randomly resulting in a total of 180 trees in each block representing every provenance. With three blocks, in total the sample trees were 540 trees.

Sampled trees were measured for their total height, clear bole height, stem diameter at breast height, stem form, branching system, and crown form. The stem form, branching system, and crown form were assessed based on field observation on typical trees and their categories as shown in Figure 1. Scoring system (1 up to 5) was utilized to assess the stem form, branching system, and crown form parameters.

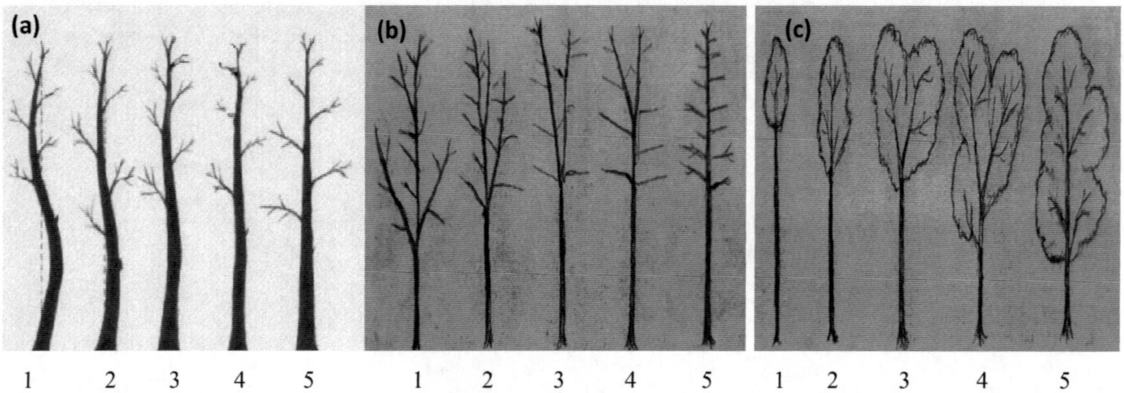

Stem forms scoring:
1 : very crooked
2 : crooked and leaning
3 : straight, crooked begin at 5 m
4 : straight, but leaning
5 : very straight

Branching system scoring:
1: forking below at 5 m
2: forking above at 5 m
3: branches at a narrow angle, and big size at the base of crown.
4: big branches at the top of crown
5: small branches with width angle

Crown from scoring:
1: very small and thin crown
2: small and thin crown, oval form
3: small crown and divided in two part with heavy branches
4: big crown with heavy branches
5: big crown with small branches

Figure 1. Categories and scoring of stem form (a), branching system (b), and crown form (c)

2.3. Measurement of Physical and Mechanical Properties

Three trees, one tree per block, were randomly selected from each provenance and cut into logs. Wood samples were taken from each log by extracting 1.5 m long log from 3 parts of stem, i.e. the bottom, middle and tip of clear bole stem. The size and sampling technique from each tree, and physical and mechanical testing methods were carried out according ASTM D.143-94: Methods of testing small clear specimen of timber [21]. Physical testing was carried out on air dry moisture content and density, while the mechanical testing was conducted on static bending (modulus of elasticity, MOE and modulus of rapture, MOR)

2.4. Statistical Analysis

The data were analyzed using randomized completely block design to compare among provenances growth traits, while for wood properties (air dry moisture content, density, modulus of elasticity, and modulus of rapture), data analysis used complete randomized design. Duncan's multiple range test at a significance level $p<0.05$ was used to compare significant differences in the means. Pearson correlation analysis was used to study the relationship among growth traits and some wood properties characteristics.

3. Result and Discussion

3.1. Provenances Growth Characteristics

The growth characteristics of *A. mangium* provenances in Parung Panjang, Bogor showed significant differences in stem diameter, clear bole height, branching system and crown form, while total height and stem form revealed not significantly different. The best growth in stem diameter was the provenance of Balimo District that reached 29.11 cm but is was not different significantly from the provenances of Bimadebun Village and Kiriwo/Serisa WP, in through out of PNG. Total height varied from 20 cm until 23 cm, clear bole height varied from 14.80 m until 18.60 m and stem form varied from 3.47 until 4.13 (Tabel 2).

Table 2. Growth characteristics of 10 provenances of *Acacia mangium*

Provenance	Stem diameter (cm)	Total tree height (m)	Clear bole height (m)	Stem form	Branching system	Crown form
Balimo District, PNG	29.11 a	22.00	17.03 bcd	3.60	4.20 a	2.00 bc
Kini WP, PNG	26.68 bc	22.93	16.33 bcd	3.47	4.13 a	1.93 bc
Lake Murray, PNG	26.07 cd	20.87	16.03 cde	4.13	3.93 b	2.87 a
Claudia River, Queensland	27.16 bcd	21.07	16.33 bcd	3.80	3.60 bc	2.07 bc
Kuru, PNG	27.01 bcd	20.33	15.90 de	4.13	3.73 abc	1.27 d
Bimadebun Village, PNG	27.95 ab	20.60	18.60 a	4.00	4.00 ab	2.53 ab
Arufi Village, PNG	27.78 b	20.83	15.80 de	3.73	3.93 b	2.60 ab
Kiriwo/Serisa WP, PNG	28.42 ab	21.33	17.60 ab	3.80	4.13 a	2.93 a
Wipim District-PNG	25.97 d	20.07	16.43 bcd	4.07	3.33 c	1.67 cd
Mata Area WP-PNG	27.23 bc	22.37	14.80 e	4.00	4.07 ab	2.03 bc
F test						
- Provenance	2.923*	0.451ns	5.737**	1.248ns	3.084*	6.081**
- Block	12.424**	0.001ns	14.040**	6.371**	172.108**	0.194ns

Notes: Different letters within a column indicate significant differences at $P < 0.05$; *Significant at $P < 0.05$; **Significant at $P > 0.01$; ns= no significant.

Branching system scoring varied from 3.33 to 4.13. The best branching system on trees is shown by the branches in the top crown [22]. In addition, the formed branches are relatively small. In the young trees, they are possibly used to self-pruning. The poor branching system is signed by the branches that are grown in the lowest position of the stem and big sizes, so it will affect the wood quality. The best performance for branching system was showed by Balimo District, Kini WP and Kiriwo/Serisa WP. Bimadebun Village provenance showed better clear bole height than others, while for branching system, Balimo District, Kini WP and Kiriwo/Serisa provenances had the highest values. The crown form which revealed the good performance was showed by Lake Muray and Kiriwo/Serisa WP provenances.

The difference in growth performance of *A. mangium* were influenced by site condition and genetic that implied in provenance itself. The tree survival and adaptability is often related with provenances (geographic) variation [23]. Genetically controlled provenance (geographic) differences are often large, especially for traits related to adaptability. The best provenance in a given site is the one that most suite to the site. Generally, provenance trials showed significant variation in height and diameter among *A. mangium* provenances as well as a significant interaction between site and provenance [24].

Some provenances revealed the best provenances based on the best performance on more the two growth parameters, such as Balimo District (stem diameter and branching system), Kiriwo/Serisa WP (stem diameter, clear bole height, branching system, and crown form), and Bimadebun Village (stem diameter, clear bole height, branching system and crown form) provenances. The characteristics of stem diameter, clear bole height, stem form and branching system affect wood quality and values. The branching system and crown form will be more useful when the tree plays as tree seed in a tree seed stand. These parameters could be implied the capability of the seed source in seed production [10].

3.2. Physical and Mechanical Properties

Effects of provenances on wood moisture content and density of *A. mangium* were significant. Wood samples from all of the trees that were sampled for this study had air moisture content varied from 15.9% (Balimo District and Bimadebun Village provenances) to 17.4% (Lake Murray provenance) (Table 3). The *A. mangium* wood moisture content in Malaysia was reported an average of 17% and 15.5% [25], while the other study showed the moisture content at 12%±2% [26]. Moisture content was known to have a great influence on strength [27].

The most essential wood properties for solid wood like sawn timber and poles (construction material) are technical quality of stems, density, strength and shrinkage [28]. The density of *A. mangium* varies depending on the provenance, ranged from 0.493 g/cm^3 to 0.559 g/cm^3 (Tabel 3). The study on *A. mangium* wood density in Sabah, Malaysia reported the ranges of wood density from 420-483 kg/m^3 based on green soaked volume [29], while the other study reported the wood density of *A. mangium* for the 8 years old ranged between 420-483 kg/m^3, and the wood density of *A. mangium* increases with age [30].

Tabel 3. Wood quality parameters of 10 *Acacia mangium* provenances

Provenance	Moisture content (%)	Density (g/cm^3)	MOE (kg/cm^2)	MOR (kg/cm^2)
Balimo District, PNG	15.9 b	0.523 ab	83825	617.5
Kini WP, PNG	16.2 b	0.496 b	82737	596.2
Lake Murray, PNG	17.4 a	0.499 b	92286	677.9
Claudia River, Queensland	17.2 a	0.502 b	86977	668.1
Kuru, PNG	17.1 ab	0.519 ab	93592	701.7
Bimadebun Village, PNG	15.9 b	0.559 a	101297	708.5
Arufi Village, PNG	17.6 a	0.512 ab	91806	645.0
Kiriwo/Serisa WP, PNG	16.9 ab	0.493 b	95616	649.6
Wipin District-PNG	17.1 ab	0.502 b	79221	606.9
Mata Area WP-PNG	16.1 b	0.521 ab	90787	648.3
F test	16.17*	2.31*	1.36^{ns}	1.84^{ns}

Notes: Different letters within a column indicate significant differences at $P < 0.05$; *Significant at $P < 0.05$; **Significant at $P > 0.01$; ns= no significant.

MOE and MOR was not affected by the provenances. The similar result was reported in Malaysia that the mean values of MOE and MOR of several provenances were not significantly different at 95% confidence level [31]. On the other hand, the effects of provenance were found to be significant for both MOE and MOR [32]. Differences of the result may be becoming from differences of sites. The MOE and MOR in this study for individual tree varied from 79221 kg/cm^2 (Wipin District provenance) to 101297 kg/cm^2 (Bimadebun Village provenance), and 596.2 kg/cm^2 (Kini WP provenance) to 708.5 kg/cm^2 (Bimadebun Village provenance), respectively (Table 3). The other study reported the MOE of *A. mangium* at 16 years old from Johore, Malaysia with MOE ranged of 77360-104210 kg/cm^2 [33].MOE and MOR are among the most important properties of the wood associated with structural quality. It is commonly used to evaluate the stiffness and strength of the wood.

3.3. Correlation among Growth Traits and Wood Properties

Simple correlation was calculated between growth parameters showed that branching system revealed positive correlation with stem diameter (r = 0.665) and total tree height (r = 0.708 (Table 4). In construction, stem form, stem diameter, and the clear bole height of wood are preferred because they are related to strength, flexibility and easiness to work. The log wood would be more efficient in use as log sawn when the stem form looked more perfect in the straightness and the height of clear bole. In this research, stem form was not significantly different among provenances and not significantly correlated with other growth traits because the stand was planted in relative dense with the spacing of 3 m x 3 m. The stem form of *A. mangium* was affected more by the environment when the trees are young than the seed origin [10].

Table 4. Inter-correlation among *Acacia mangium* growth traits

	Stem diameter (cm)	Total tree height (m)	Clear bole height (m)	Stem form	Branching system
Total tree height (m)	0.369				
Clear bole height (m)	-0.248	0.442			
Stem form	-0.393	0.496	-0.300		
Branching system	0.665*	0.708*	0.045	-0.508	
Crown form	0.321	0.221	0.120	0.011	0.510

Note: **Significant at P < 0.01

Some growth parameters have had significant correlation with wood properties. The total tree height, stem form, and branching system were revealed significant correlation with MOR, with coefficient of correlation -0.678, 0,719 and -0.785, respectively (Table 5). Modulus of Rupture, frequently abbreviated as MOR, (sometimes referred to as bending strength), is a measure of a specimen's strength before rupture. It can be used to determine a wood species' overall strength.

Table 5. Correlation among *Acacia mangium* growth traits and wood properties

	Moisture content	Wood density	MOE	MOR
Stem diameter	-0.136	-0.206	-0.388	-0.538
Total tree height	-0.466	-0.361	-0.476	-0.687*
Clear bole height	-0.090	-0.407	-0.065	-0.219
Stem form	0.254	0.327	0.485	0.719*
Branching system	-0.054	-0.612	-0.559	-0.785**
Crown form	0.298	-0.489	-0.069	-0.314

Note: * Significant at P < 0.05

Most of growth parameters was not significantly correlated with physical and mechanical wood properties. This result indicated that the most of growth characteristics were not a good indicator for selecting the physical and mechanical wood properties of *A. mangium*, except for total tree height, stem form, and branching system. Similar study was reported on *Eucalyptus camaldulensis,* that stated there were no significant differences in basic, oven-dry and green density values between fast and slow growth [15]. The other study also stated that not significant association was found between wood basic density and growth and form traits of *Pinus ponderosa*, except for stem straightness [34]. Some studies also reported that growth rate differences have little effect on wood property of hardwoods [16,17]. The provenance that growing faster compared to other provenances does not mean will affect the wood physical and mechanical traits [12], and there is no consistent relationship was found between wood physical and mechanical traits and growth rate in diffuse-porous hardwoods [35].

Most of the correlation was a negative, except for stem form. The negative correlation indicated that increase of growth caused the decrease of wood quality. The genetic correlations between stem straightness and growth traits in the progeny test of *A. mangiun* in Malaysia were favourable but weak, while those between growth traits and dynamic modulus of elasticity (MoE) were weak and unfavourable [36]. The result was similar with the study on *Pinus radiata* [37]. The other studies also reported a weak genetic control in the stem form (tendency for forking and in straightness of the bole) for A. *mangium* provenances [7,38]. The substantial difference in the forking of *A. mangium* was found among provenances, but an even larger difference between different locations among some provenances [6].

3.4. Implication for Provenances Seed Stand

For solid wood like sawn timber and poles (construction material) the most essential wood properties are technical quality of stems, density, strength and shrinkage [28]. In construction, straight stem or wood is preferred because straightness is related to strength, flexibility and easiness to work. Stem straightness is a big factor affecting harvesting efficiency [39]. The best performance of *A. mangium* based on growth traits was Kiriwo/Serisa WP Bimadebun Village, and Balimo District, but from wood physical and mechanical traits, the superior provenance was revealed by Bimadebun Village with the lower moisture content, the highest wood density, MOE and MOR. The provenance is potential to be developed as provenance seed stand, especialy for solid (contruction) wood.

4. Conclusion

Provenance difference was significant affected on growth performance of *A. mangium,* except for total height and stem form. Most of growth parameters was not significantly correlated with wood physical and mechanical traits, except for MOR, it revealed significant correlation with the total tree height (r = -0.678), stem form (r = 0.719), and branching system (-0.785). Generally, some growth characteristics were not a good indicator for selecting the best provenance in physical and mechanical wood properties of *A. mangium.* The best performance of *A. mangium* based on growth traits was Kiriwo/Serisa WP Bimadebun Village, and Balimo District, but from wood physical and mechanical traits, the superior provenance was revealed by Bimadebun Village with the lower moisture content, the highest wood density, the highest MOE and MOR. The provenance is potential to be developed as provenance seed stand.

Acknowledgement

This study was funded by the Forest Tree Seed Technology Research and Development Institute. The authors are grateful to Nurwati Hadjib for helping in the wood quality testing and the operational staffs in Parung Panjang Forest Research Station who helped in field data collection.

References

[1] Lee S S 2004 *Unasylva* **287(55)** 31-35
[2] Hardiyanto E B 2014 Challenges for Acacia breeders *Sustaining the future of Acacia plantation forestry* ed Hardiyanto E B (Hue Vietnam: IUFRO Mar 18–21, 2014)
[3] Ministry of Forestry 2013 *Statistic of Indonesia forestry* (Jakarta: Ministry of Forestry, Republic of Indonesia)
[4] Hegde M, Palanisamy K and Yi J S 2013 *J For Sci* **29(1)** 1-14

[5] Krisnawati H, Kallio M and Kanninen M 2011 *Acacia mangium Willd: ecology, silviculture and productivity* (Bogor, Indonesia: CIFOR)
[6] Awang K and Bhuimibhanon P 1993 Genetics and tree improvement *Acacia mangium, Growing and Utilization* eds Awang K and Taylor D (Bangkok: Winrock International and the Food and Agriculture Organization of United Nations)
[7] Harwood C E, Aplegate G, Robson K and Williams A R 1993 Early growth of provenances and progenies in *Acacia mangium* seed production areas in North Queensland, Australia. Acacias for Rural, Industrial and Environmental Development eds Awang K and Taylor D (Bangkok: FAO and Winrock International)
[8] Otsamo R 1996. Integrating indigenous tree species in forest plantations on *Imperata cylindrica* grasslands-results from Riam Kiwa, South Kalimantan *Reforestation: Meeting the Future Industrial Wood Demand. Proceedings of a Workshop held in Jakarta, 30 April-1 May 1996* eds Otsamo A, Kuusipalo J and Jaskari H (Jakarta: Ministry of Forestry of Indonesia)
[9] Tuomela K, Otsamo A, Kuusipalo J, Vuokko R and Nikles G 1996 *For Ecol Manag* **84** 241–249.
[10] Nurhasybi, Sudrajat D J and Diatna K 2009 *Indonesian J For Res* **6(1)** 1-16
[11] Nurwati, Hadi Y S nad Setyaningsih D 2007 *JITKT* **5(1)** 7-11
[12] Zobel B J and van Buijtenen J P 1989 *Wood variation its causes and control* (Berlin, Heidelberg: Springer. Series in Wood Science)
[13] Saranpää P 2003. Wood density and growth. *Wood quality and its biological basis* eds Barnett J R and Jeronimidis G (Blackwell Publishing and CRC Press. Biological Sciences Series)
[14] Moya M R and Tomazello M 2007 *Investigación Agraria: Sistemasy Recursos Forestales* **16(3)** 267-276
[15] Malik M F E I and Abdelgadir A Y 2015 *J For Prod Indus* **4(3)** 86-93
[16] Jayawardana D N and Amarasekara H S 2008 Investigation of the effect of growth rate on the quality of Teak (*Tectona grandis*) wood *Proceedings of the International Forestry and Environment Symposium* 2008 (Kalutara, Sri Lanka)
[17] Kojima M, Hiroyuki Y, Kayo O, Yasuhisa O, Masato Y, Takashi O, Toshihiro O, Kenji M, Kentaro N and Yuji I 2009 *J Wood Sci* **55(6)** 417-424
[18] Jiang J H, Wang XQ, Fei B H, Ren H Q and Liu XE 2007 *Ann For Sci* **64** 807-814.
[19] Pande P K 2010 *American J Plant Sci* **2** 644-649
[20] Sudrajat D J 2016 *For Sci Tech* **12(1)** 9-15
[21] ASTM (American Society for Testing and Material) 2005 *Annual Book of ASTM Standards. Section Four: Construction.* Volume 0410 Wood D-143–94 (Reapproved 2000) Standard Test Methods for Small Clear Specimens of Timber (West Conshohocken USA: American Society for Testing and Material).
[22] Laurisen E B, Kjaer E D and Nissen M 1995 *Second evaluation of an international series of Gmelina arborea provenances trials* (Humlebaek-Denmark: Danida Forest Seed Centre)
[23] Zobel B J and Talbert J 1984 *Applied forest tree improvement* (Ilinois: John Wiley and Sons, Inc.)
[24] Lokmal N and Ernest C O K 1995 *J Trop For Sci* ***8*** 247-254
[25] Peh T B and Khoo K C 1984 *Malaysian For* **47** 285-303
[26] Lokmal N and Abdul-Ghani M N 2014 *J Agric Res Dev* **5(5)** 079-084
[27] Madsen B 1975 *Canadian J Civil Engin* ***2*****(4),** 466-473
[28] Laurila R 1995 *J Trop Fort Prod* **1** 209-221
[29] Razali A K and Mohd H S 1993 Property and utilization. *Acacia mangium - Growing and Utilization. MPTS Monograph Series No. 3* (Bangkok, Thailand: Winrock International and FAO)

[30] Lim S C, Gan K S and Tan Y E 2011 *Properties of Acacia mangium planted in Peninsular Malaysia ITTO Project No. PD 306/04(1)* eds Lim S C, Gan K S and Tan Y E (Kepong: Forest Research Institute Malaysia)
[31] Sahri M H, Ashaari Z and Kaderi R A 1998 *Pertanika J Trap Agric Sci* **21(2)** 73 – 81
[32] Ngah L and Abdul-Ghani M N 2014 *J Agric Res Dev* **5(5)** 079-084
[33] Omar M K M and Jamil A W M 2011 Mechanical properties. *Properties of Acacia mangium planted in Peninsular Malaysia ITTO Project No. PD 306/04(1)* eds Lim S C Gan K S and Tan Y E (Kepong: Forest Research Institute Malaysia)
[34] Martinez-Meier A, Gallo L, Pastorino M, Mondino V and Rozenberg P 2011 BOSQUE **32(3)** 221-226
[35] Bowyer J L, Shmulsky R and Haygreen J G 2003 *Forest products and wood science – An introduction* Fourth edition (Iowa USA: Iowa State Press)
[36] Hai P H, Duong L A, Toan N Q and Ha T T T 2015 *New For* **46** 577 - 591
[37] Gapare W J, Baltunis B S, Ivkovi M and Wu H X 2009 *Ann Fort Sci* **66** 1-9
[38] Nirsatmanto A, Leksono B, Kurinobu S and Shiraishi S 2004 *J For Res* **8** 127-131
[39] Davis L S 1970 *Economics models for program evaluation* Paper FO-FTB-69 13/2 Second World Consultation on Forest Tree Breeding (Rome: Food and Agricultural Organization) p 226

Pirdot (*Saurauia bracteosa* DC) Leaf Processing Technique for Making Herbal Tea

RO Situmorang* **and AD Sunandar****

Environment and Forestry Research and Development Institute of Aek Nauli, Simalungun, Indonesia.

email: *pita_80s@yahoo.com; **sunandar_dany@yahoo.com

Abstract. Pirdot is one of forest plants that have the potential to be made into herbal tea. This study aims to obtain the best processing techniques and herbal tea formulas for processing pirdot leaves into herbal tea in the form of tea bags. The processing of pirdot herbal tea used three or five fresh pirdot leaves count from the shoots. Leaves processing was held in two ways, through a fermentation and without fermentation process. Organoleptic test of 15 panelists were carried out to test tea color, aroma and taste. Data were analyzed by applying Kendal's tau multilevel correlation test. The results showed that the mixture of the top five leaves gave better result in color, aroma and taste of pirdot tea. Fermentation process also gave better results in terms of color, taste and aroma of pirdot tea compare to no fermented tea. However, optimum condition in fermentation process is still being studied.

1. Introduction

Tea is a beverage originating from various types of plant leaves, especially tea plants and a common beverage. In its development, tea is not only derived from the processing of tea plants but also other various types of plants. According to [1] tea can be grouped into two classes, namely herbal tea and non-herbal tea. Non-herbal teas are grouped into three: black tea, green tea and oolong tea. Herbal tea is the result of processing of flowers, bark seeds, leaves and roots of various plants. One of the plant that has the potential to be processed for herbal tea is Pirdot (*Saurauia bracteosa* DC).

Pirdot is one of the endemic plants that grows in various places in Indonesia and is used primarily by the Bataknese as a medicine. In North Sumatera Province, this tree is found in Simalungun Regency in the north to Tapanuli Selatan in the south especially in the highland. Even though it spreads in many places, this species is categorized as vulnerable according to IUCN (1998) and [2] classified pirdot in critical categories. Bataknese, especially in Toba, use pirdot as a traditional medicine to reduce blood sugar levels, prevent high blood pressure and reduce blood cholesterol levels. Pirdot is a pioneer tree species that grows in open area, along rivers and in humid places, at altitudes of 600 - 1,200 meter above sea level [3]. This species is categorized as small-medium tree species with a height between 3-15 meters [3]. The trunk is easily broken and because it is relatively small so it is not widely used by the community. The spread occurs naturally, the propagation and cultivation knowledge is still unknown.The benefits of pirdot in the world of health have been widely carried out, among others, as anti-oxidants [4] [5] [6]. This study aims to obtain the best techniques in processing pirdot leaves into herbal tea in the form of tea bags.

2. Materials and method

2.1. Materials

The materials used in this study was fresh pirdot leaves that collected from arboretum of Environmental and Forestry Research and Development Institute of Aek Nauli. The equipment used were tea bags, sealers, incubation chambers, thermometers, hygrometers, ovens, digital scales, storage containers, knives, checks and gloves.

2.2. Research Design

First Step. Determination of pirdot processing techniques

At this stage there are two basic treatments, namely material differences and manufacturing methods differences. The different age of the leaves was considered affecting the quality of the leaves from their chemical content, as well as the processing of herbal medicines [8]. The difference in treatment of different raw materials was needed to find out the best raw materials used in processing pirdot herbal tea. In this case there were two different types of raw materials:

- The top 5 pirdot leaves (top shoots were counted from above)
- The top 3 pirdot leaves.

Afterwards, treatment processing was performed. There were two categories of thetreatment processing:

- No enzymatic oxidation (no fermented)
- Enzymatic oxidation (fermented)

The extend treatment levels were as follows:

- P1 = top 5 pirdot leaves were not oxidized (no fermented)
- P2 = top 3 pirdot leaves were not oxidized (no fermented)
- P3 = top 5 pirdot leaves were oxidized (fermented)
- P4 = top 3 pirdot leaves were oxidized (fermented)

Figure 1. Pirdot leaves as raw material.

The procedure of tea processing was as follows:

- Fresh pirdot leaves were collected and sorted
- For non-fermented treatment, the leaves were sliced 1-2 cm wide, then dried for 4 days in room temperature (until water content 10-15%)
- Dried leaves were mashed by coarse tendering, and ready to further process.
- For fermentation treatment, fresh leaves were left wilted (1-2 days, depending on the room temperature and moisture content). Wilted leaves were recognized by squeezing the leaves whether they have felt wilted or still break easily.
- Wilted leaves were twisted (rolled), kneaded, and/or sliced into a small size and then spread on fermentation (oxidation) slabs with 1-2cm thickness.
- Sliced leaves were then incubated in the incubator room in controlled temperature, humidity for 6 hours
- Fermented leaves were dried until reach 8-10% moisture content 8-10%
- Dried leaves were put into tea bag and tested

In the fermentation process of pirdot leaves, the condition of the processing room could be modified to get the most ideal humidity and temperature. Modification was made based on tea leaves fermentation which humidity (RH) is above 90% and temperatures is ranging from 23-27°C. Moisture could be added by fading the chamber if it is too low. Oxidation was carried out for 6 hours.

2.3 Data Analysis

The characteristics of tea were determined by powder size, powder color, steeping water color, aroma and taste [9]. Organoleptic test was performed by comparing the results of each treatments. The testing parameters were color, aroma, and taste of the steeping tea. The grading of scores was Likert scale, were as follows:
1 = dislike extremely
2 = dislike (very much)
3 = neither like or dislike = like
5 = like very much

Data obtained from the first step testing (treatment of raw material differences and processing processes) were tested with a multilevel correlation (*Kendal Tau*) using SPSS 23.0. Testing *Kendal Tau* was chosen because the type of data produced was ordinal data, and the parameters observed do not have a level (in the form of a treatment group).

3. Result and discussion

Polyphenol or phenolic compounds are bioactive components that have antioxidant activity which is naturally found in vegetables and fruits such as tea (tree/plant/shrub). Polyphenolic compounds consist of several subclasses, such as flavonol, flavone, antocianidine, catechin and biflavan [10]. Phenolic compounds of flavonoids are also found in pirdot leaves. Additionally, pirdot leaves also contains secondary metabolites such as alkaloids, steroids, and tannins [7].

Tea is rich in polyphenolic flavonoids which have shown strong antioxidant potential and have been tested in vitro and in vivo. The content of polyphenolic flavonoids in tea depends on the type of raw material and the processing method [11]. Herbal tea is one of tea and herbal drink products that helps in health treatment [12].

Pirdot leaves are commonly consumed in the form of tea. Pirdot leaves are widely used, especially in North Sumatra for its health benefit. Pirdot leaves are also consumed as stamina booster. Based on a survey conducted by several respondents around Toba Lake (Districts of Simalungun and Toba Samosir), it was stated that several diseases can be cured by consuming pirdot leaves. Some of them were lymph node cancer, thyroid cancer, uterine tumors, high blood pressure, diabetes, gout, and various open wounds due to itching, boils, cut wounds that occur in humans or livestock [13]. However, it needs laboratory-scale testing both in-vitro and in-vivo tests to prove the efficacy of pirdot leaf as a medicine. This study focused on the processing of pirdot leaves, thus its utilization and added value can be improved.

During this time, the use of pirdot leaves by the community is still very simple through boiling. The trial of processing pirdot herbal tea is the first step to better utilize pirdot leaves for wider consumers. The processing of this traditional method needs to be improved so that the presentation is easier, the dosage is more precise, the appearance is more attractive, and easily distributed. Efforts to improve primary processing technology are carried out by making pirdot tea bag. [14] argued that the transformation of the supply of herbal tea from traditional methods to more modern, supervised, and standardized processing techniques was to increase the added value, to prevent contamination during the processing, and to provide a more appropriate presentation and dosage.

In processing herbal tea, the quality of tea is greatly influenced by the level of aging of the leaves and the method of processing. Leaf composition affects the content and type of polyphenols [15]. In tea leaves (*Camellia sinensis*), levels of young leaf polyphenols are higher than the levels of old one [16]. On the other hand, [17] found that old breadfruit leaves have higher levels of polyphenols than young leaves. Based on these results, polyphenol content varies in young and old leaves, thus it could not be standardized that the older the leaves, the lower the content of the polyphenols and vice versa.

Differences in raw material sources and processing techniques are also carried out in the manufacture of pirdot herbal tea. The results of organoleptic testing of different sources of raw materials and tea processing techniques can be seen in Table 1.:

Table 1. Results of organoleptic test of differences of raw material sources and tea processing techniques.

Treatment	Color	Aroma	Taste	Average
P1	1.80	2.27	2.60	2.22
P2	1.33	2.47	2.47	2.09
P3	2.87	4.07	3.80	3.58
P4	2.40	3.13	3.27	2.93
Average	2.10	2.98	3.03	

Note: Total respondent =15 panelists

The results in Table 1 descriptively show that there were differences in color quality, aroma, and taste of pirdot herbal tea processed from leaves from the top 5 and 3 shoots, as well as oxidation process. Leaves obtained from the top 5 pirdot leaves were more prefer for its color, aroma and taste (P1 and P3). Furthermore, oxidation activities improved the value of the respondent's preference for the tested parameters.

Figure 2. Pirdot tea powder.

3.1. Color

The results of brewing pirdot herbal tea appear to be disliked by respondents (average score of 2.10). Pirdot tea steeping showed that the color of steeping water does not look bright when compared to other teas, such as green tea, jasmine tea, and others. Tea is identical to the color of steeping which has a bright color adjusted to its base material. The color of pirdot herbal tea tends to be less bright or less attractive, which is likely to have low phenolic leaves. Further research to identify the phenolic types and to test quantitative phenolic levels in pirdot leaves is indispensable. Although colors are less bright than commonly consumed tea, different types of raw materials and processing techniques gave different results. The results of tea pirdot steeping with four types of treatment were shown in Figure 3.

Figure 3. Results of brewing pirdot herbal tea.

From Table 1, it can be seen that the best color quality of pirdot leaves is obtained from a combination of older age leaves and oxidation process. From these results it could be seen that the content of phenolic compounds of leaves derived from older leaves is higher when compared to leaves from young shoots. To prove this result, laboratory tests are still needed, such as testing phenolic content and antioxidants in herbal tea powder. The results in steeping water quality from leaves without oxidation (no fermented) and with oxidation (fermented) were also different where fermented leaves produces darker color. According to [18] the enzymatic oxidation process contained is responsible for color changes in various plant tissues containing phenolic. This term is known as enzymatic browning. The difference in the color of the pirdot herbal tea made by oxidation is due to the fact that during the oxidation process, there is an enzymatic process that produces brownish color to the food.

Correlation test (*Kendal tau* test) was carried out to test the significance relationship between parameters of leaf aging and oxidation treatment against respondent's preference. Result of *Kendal tau* correlation can be seen in Table 2.

Table 2 showed that there were very significant differences in processing techniques for color quality that applied in treatment P1 and P3 (with oxidation and not oxidized). Nevertheless, differences in raw materials (level of aging) had no significance correlation in the treatment. These results indicated that the treatment of processing techniques greatly affects the color of tea. The difference in color of pirdot herbal tea produced by oxidation process was due to the fact that during the oxidation process there was an enzymatic process which produces brown color on the wet tea powder. Enzymatic browning occurs in foodstuffs including leaves that contain many phenolic substrates. This reaction occurs if plant tissue is cut, peeled and due to mechanical damage that can cause damage to plant tissue integrity. This causes enzymes come into contact with the substrate which is usually an amino acid tyrosine and phenolic components such as catechins, caffeic acid, and chlorogene acid. That process made phenolic substrate in the plant then hydroxylated to 3,4-dihydroxyphenylalanine (dopa) and oxidized to quinone by phenolase enzymes causing the color of food ingredients to become brownish [18]. During oxidation process, the oxidation chamber was conditioned so that temperature and humidity were adjusted to the optimum activity of oxidizing enzymes (phenolase). Water content during the oxidation process was also considered to greatly affect the quality of tea color and provide uniformity of tea color [19].

Table 2. Correlation of leaves aging and oxidation process on pirdot tea color.

			P1 Color	P2 Color	P3 Color	P4 Color
Kendall's tau_b	P1 color	Correlation Coefficient	1.000	-0.020	0.691(**)	0.222
		Sig. (2-tailed)	.	0.938	0.006	0.382
	P2 color		-0.020	1.000	0.408	0.318
			0.938	.	0.104	0.212
	P3 color		0.691(**)	0.408	1.000	0.380
			0.006	0.104	.	0.129
	P4 color		0.222	0.318	0.380	1.000
			0.382	0.212	0.129	.

Note : **Correlation is significant at 0.01 level (2-tailed). N = 15

In line with the results of fermentation treatment in other studies that provide significant results in the color of pirdot tea steeping, research conducted by [20] also produces results that are not different. In his research on the manufacture of herbal teas which are processed from chrysanthemum flowers, fermentation process resulted stronger color in steeping water.

3.2. Aroma

The average value of panelist preference for the aroma of pirdot herbal tea (Table 1) looks better compared to color preference (average value of 2.98). From these results, oxidized pirdot leaves were more preferred in aroma rather than the nonoxidized. The difference in leave ageing did not show linear results. A correlation test result of the treatment could be seen in Table 3.

Table 3. Correlation of leaves aging and oxidation process on pirdot tea aroma.

			P1 Aroma	P2 Aroma	P3 Aroma	P4 Aroma
Kendall's tau_b	Aroma P1	Correlation Coefficient	1.000	-0.262	0.587(*)	0.132
		Sig. (2-tailed)	.	0.327	0.025	0.613
	Aroma P2		-0.262	1.000	0.152	0.546(*)
			0.327	.	0.562	0.036
	Aroma P3		0.587(*)	0.152	1.000	0.521(*)
			0.025	0.562	.	0.042
	Aroma P4		0.132	0.546(*)	0.521(*)	1.000
			0.613	0.036	0.042	.

Note : **Correlation is significant at 0.05 level (2-tailed).* N = 15

The correlation between oxidation treatment and leaves aging provide significant results in the aroma of pirdot tea steeping. This could be seen from the difference in treatment without oxidation and after oxidation in treatment P1 and P3, and between P2 and P4. Furthermore, the treatment is oxidized on raw materials which have different levels of aging (P3 and P4) also provide significant results. The content of plants that provide aroma to herbal plants is due to the presence of volatile compounds that easily reach the sense of smell and when there is sufficient concentration of aroma compounds in the plant [21]. The GC-MS testing in the earlier study [13] showed that there were at least 20 types of chemical compounds found in pirdot leaves. However, the classification of aroma and volatile compounds in pirdot leaves have not been carried out.

Non-oxidized treatment (P1 and P2) result showed that there was no significant difference in the in leaves aging level to the tea aroma. Nevertheless there were significant differences (level 0.05) to the tea aroma after oxidation treatment (P3 and P4). Based on these results, it can be seen that there were differences in aroma triggered from different leaves age, where the older leaves have better aroma quality compared to the younger leaves. The chemical content in plant leaves is a scent or an aroma of herbal tea. Oxidation activities such as drying, roasting, or fermentation and other activities are carried out to reduce enzymes along with reduced water, temperature, or other activities to obtain a better aroma of herbal tea.

Enzymatic oxidation treatment in pirdot herbal tea manufacture yielded significant results of the two types of leaves (raw materials) tested. Oxidation treatment omitted raw odor which is caused bad smell in herbal teas. Unfermented tea has raw taste and aroma like green grass because there are still many enzymes that cause odor [19]. The process of oxidation (fermentation) activities such as withering and grinding results the trapped enzymes and the odor in cells to be broken down and evaporate during the grinding process, and become non-active due to a decrease in moisture in the drying. The leaf grinding process, in addition to extracting enzymes, also functions to extract the chemical content (source of aroma) contained in the leaves so that the aroma and flavor becomes increased. Furthermore, in fermentation process, the enzymatic activity of phenolase compounds also inhibits the activity of other enzymes in obtaining substrate. Considering this process, the combination of activities in the oxidation process such as withering, rolling, fermentation, and drying resulted in increased aroma through the extraction of aroma-producing chemical compounds and non-active enzymes that cause the raw odor in pirdot herbal tea.

3.3. Taste

The average value of panelist preference for the taste of herbal pirdot tea (Table 1) was 3.03. This results showed that tea obtained from top 5 leaves was preferred compared to tea from top 3 leaves and the oxidized tea was preferred with non-oxidized ones.

The average panelist score showed the level of panelist preference is still low (3 = less preferred). This is considered to be normal, because pirdot tea is an herbal tea (medicine) which has a more bitter taste compared to other ordinary well known teas. According to panelists, pirdot tea has a rather bitter taste and is dominated by a feeling of tightness on the tongue. To achieve maximum tea flavor, pirdot tea should be processed into finer powder so it can increase the absorption of color, taste, and aroma in steeping water. Table 4 showed the results of correlation test to find out the differences in the treatment of leaves aging and oxidation treatment on taste quality of pirdot herbal tea.

Table 4. Correlation of leaves aging and oxidation process on pirdot tea taste.

			P1 Taste	P2 Taste	P3 Taste	P4 Taste
Kendall's tau_b	P1 Taste	Correlation Coefficient	1.000	0.842(**)	0.372	0.672(**)
		Sig. (2-tailed)	.	0.001	0.136	0.008
	P2 Taste		0.842(**)	1.000	0.247	0.731(**)
			0.001	.	0.336	0.005
	P3 Taste		0.372	0.247	1.000	0.641(*)
			0.136	0.336	.	0.010
	P4 Taste		0.672(**)	0.731(**)	0.641(*)	1.000
			0.008	0.005	0.010	.

Differences in the aging rate of leaves affected on taste values (treatment P1 and P2 and treatment of P3 and P4) significantly and fermentation treatment also had a very significant effect on taste quality (P2 and P4 treatment). [22] indicated that catechins (major polyphenols) are the main component of the causes of bitter taste in medicinal products (plants). Besides, caffeine, saponins [23] and alkaloids [24] also contribute to the bitter causes. [25] added that catechins are an important component of tea and contributes to color, pungency and taste characteristics. One function of fermentation and oxidation in tea is to improve taste and aroma. Sufficiently unfermented tea has a raw taste and aroma or taste and aroma like grass [19]. Herbal tea produced from older pirdot leaves has a better taste value compared to leaves obtained from younger one. However, it should be noted that there is an age limit of leaf to be used in tea production and very old leaves are usually less qualified and the tissue is too hard. In this experiment the leaf boundaries used were five leaves from the shoot.

Furthermore, in the treatment of enzymatic oxidation, there was a forging activity and destruction of plant tissue which is carried out slowly. This resulted in the extraction of chemical compounds that cause taste. The existence of a combination of activities in the oxidation process such as withering, rolling (milling), fermentation, and drying resulted in an increase in taste through the extraction of chemical compounds that produce aroma and non-active enzymes that cause odors and raw flavors.

4. Conclusions and recommendations

The differences in the level of leaves aging and oxidation treatment had a significant effect on the quality of color, aroma, and taste of pirdot herbal tea. Herbal tea obtained from the top five leaves showed better quality of color, aroma and taste compared to tea obtained from the top three leaves. This indicated that the chemical content in older pirdot leaves such as flavonoid compounds, alkaloids, or other chemical compounds was higher than the younger ones. Oxidizing enzymatic treatment (fermentation) in the manufacture of herbal tea pirdot provide a better quality in color, aroma, and taste. The milling process (tissue destruction) and oxidation were conducted in controlled temperature, humidity, and time to support the enzymatic browning that can enhance the color, aroma, and taste of herbal tea. Enhance information about chemical content of pirdot leaf such as the content of tannin, anti-oxidants and other chemical compounds in the processed tea is required.

Acknowledgment

Authors wishing to acknowledge Environment and Forestry Research and Development Institute of Aek Nauli for financial support and also Selamat Lumban Tobing and Firman as technician staff for all technical support.

Reference

[1] IRD A. Indonesia tea catechin: prospect and benefit. J Kultiv. 2016;15(2):99–106.

[2] Wihermanto. Dispersi Asosiasi dan Status Populasi Tumbuhan Terancam Punah di Zona Submontana dan Montana TN Gunung Gede Pangrango. Biodiversitas. 2004;5(1):17–22.

[3] Steenis C van. Flora Pegunungan Jawa. Bogor: Pusat Penelitian Biologi-LIPI; 2010.

[4] Yonagarasirham S, Chelladurai V, Togunashi V, Govindaiah, Murthy K. Saurauia brecteosa DC. (Sauriaceae) – a new record for the Indian subcontinent and a new generic record for Andaman & Nicobar Island. Curr Sci. 1982;51(4):198–9.

[5] Maukar M, Runtuwenne M, Pontoh J. Analisis kandungan fitokimia dari uji toksisitas ekstrak methanol daun soyogik (Saurauia bracteosa DC) dengan menggunakan metode maserasi. J. Ilm Sains. 13AD;2(98–101).

[6] Kadji M, Runtuwenne M, Citraningtyas G. Uji Fitokimia dan Aktivitas Antioksidan dari Ekstrak Metanol Daun Soyogik (Saurauia brectiosa DC). Pharmacon. 2013;5:13–7.

[7] Situmorang R, Harianja A, Silalahi J. Karo's Local Wisdom: Teh Use of Woody Plants For Traditional Diabetic Medicines. Indones J For Res. 2015;2(2):121–31.

[8] Pothinuch S, Tongchitpakdee. Melatonin contents in mulberry (Morus spp.) leaves: Effects of sample preparation, cultivar, leaf age and tea processing. Food Chem. 2011;128:415–9.

[9] Lee J, Chambers D. A Lexicon for flavour description analysis of green tea. J Sens Stud. 2007;22:256–72.

[10] Astawan M. Tetap Sehat Dengan Produk Makanan Olahan. Surakarta: PT Tiga Serangkai; 2004.

[11] Toda S. Polyphenol Content and Antioxidant Effects in Herb Teas. Chin Med. 2011;2:29–31.

[12] Hambali E, Nasution M, Herlina E. Membuat aneka herbal tea. Jakarta: Penabar Swadaya; 2005.

[13] Sunandar A, Situmorang R. Perbanyakan vegetatif, pengolahan sederhana dan pembuatan demplot pirdot. 2016.

[14] Chan K. Some aspects of toxic contaminants in herbal medicines. Chemosphere. 2003;52:1361–71.

[15] Felicia N, Widarta W, Yusasrini N. Pengaruh Ketuaan Daun Dan Metode Pengolahan Terhadap Aktivitas Antioksidan Dan Karakteristik Sensoris Teh Herbal Bubuk Daun Alpukat (Persea americana Mill.). Universitas Udayana; 2016.

[16] Izzreen N, Fadzelly. Phytochemicals and Antioxidant Properties of Different Parts of Camellia sinensis leaves from Sabah Tea Plantation in Sabah, Malaysia. IFJR. 2013;20(1):307–12.

[17] Mu'nisa A, Pagarra H, Muflihunna S. Uji Kapasitas Antioksidan Ekstrak Daun Sukun dan Flavanoid. Univ. Negeri Makassar; 2011.

[18] Simpsons B, editor. Food Biochemistry and Food Processing. 2nd ed. New Yor: Willey-Blackwell; 2012.

[19] Toit J, Jubert E. The effect of pretreatment on teh fermentation of honeybush tea (Cyclopia mucultata). J Sci Food Agric. 1998;76:537–45.

[20] Sudarmanto. Pengaruh Fermentasi Dua Varietas Bunga Krisan (Chrysantehmum Sp) terhadap Mutu Teh Krisan. Jakarta; 2003.

[21] Antara N, Wartini M. Senyawa Aroma dan Cita Rasa (Aroma and Flavor Compounds). Denpasar; 2015.

[22] ZM C, HF W, XQ Y, N X. The chemistry of tea non-volatiles. In: Y Z, editor. Tea Bioativity and Therapeutic Potential. New York: Taylor and Francis; 2002. p. 57–88.

[23] Wang L, Kim D, Lee C. Effects of heat processing and storage on flavanols and sensory qualities of green tea beverage. J Agric Food Chem. 2000;48:4227–32.

[24] RP D. Pengaruh Perbandingan Teh Herbal Daun Binahong (Anredera cordifolia (Ten.) Steenis) dengan Daun Teh (Camellia sinensis) dan Suhu Pengeringan terhadap Karakteristik Teh Herbal. Bandung; 2017.

[25] Harler C. Tea manufacture. New York: Oxford University Press; 1963.

IOP Conf. Series: Earth and Environmental Science **359** (2019) 012005 doi:10.1088/1755-1315/359/1/012005

Influence of Site on Streak Patterns and Several Physical Properties of *Diospyros celebica* Bakh.

M Asdar
Environment and Forestry Research and Development Institute of Makassar.
Jl. Perintis Kemerdekaan Km. 16 Makassar, Indonesia

email: muh.asdar@gmail.com

Abstract. Ebony has a beautiful heartwood with black and brown streaks alternately. Each site has various streak pattern in which the industry classified it into two streak types namely *batang macis* (BM) and *sarang laba-laba/ bendera* (SL). This research was conducted to investigate the influence of site on streak patterns and several physical properties of the heartwood of ebony. Ebony from Parigi has BM streaks pattern and meet the grade B-A, meanwhile ebony from Poso has SL streaks pattern, qualified for grade C only. On the basis of t-test, only L* (lightness-darkness) was differed between the two streak patterns, while wood density and initial moisture content were not differed significantly. The BM streaks pattern has a darker of heartwood color, higher density but lower initial moisture content.

1. Introduction

Ebony (*Diospyros celebica* Bakh.) is an endemic species of Sulawesi [1]. Ebony has a beautiful heartwood, displayed series of axial and alternating black and brown streakes. The streak pattern determines the quality of ebony especially for exports to Asian countries.

Ebony exporters in South Sulawesi and Central Sulawesi have their own standards related to the streak pattern. Thin and parallel black streaks with almost the similar interval are considered to be the highest quality. In contrast, wide black streaks, varying thicknesses and irregular intervals have lower quality. In the realm of ebony industry, there is a type of *batang macis* (BM, matchstick), an ebony with thin, similar of width, and parallel black streak and *sarang laba-laba* (SL, spider web) for spaced and disorganized streak. Standar Nasional Indonesia 01-2028-1990 concerning processed ebony wood [2], does not specify the requirements of wood streaks but only requires straight and parallel streaks in each sortimen. Ebony wood is exported in the form of chopsticks, butsudan and boards sortimen.

The streak pattern is thought to be caused by site. Ebony in natural habitat produces different streaks [3-6]. Ebony BM has only been obtained in Donggala and Parigi Moutong Districts, Central Sulawesi. Outside these areas generally produce SL streaks. Previous researchers [7-8] presume that the width and color of the lines/streaks and the shape of the heartwood in the stems may be influenced by the site.

The influence of the site on streak patterns and the physical properties of the heartwood of ebony are unknown.Hence, the purpose of this study was to observe streak patterns, heartwood color, initial moisture content and the density of ebony BM and SL heartwood that grows at different sites.

2. Research Methods

Ebony BM samples were taken from Sausu, Parigi Moutong District, coordinates 01° 04 '055 "S. latitude 120° 31' 687" E longitude and ebony SL in Towu, Poso District at 01° 19 '109 " S latitude 120° 36' 354" E longitude, Central Sulawesi Province. Two trees were collected from each site with an average diameter of 40 cm.

Wood samples were taken at the base of the tree in the form of disks as thick as 5 cm. The research procedure was as follows:

- To observe the streak pattern, the disks were sawn in quarter sawn trough the pith. In this study, the determination of quality is only focused on the streak pattern, while the length, width and thickness of the specimen are ignored. Determination of the quality of the wood streak is based on the standards used by the ebony wood industry in Palu and Makassar. Quality A and B are classified as ebony BM. The standard used is based on visual observations (Figure 1).

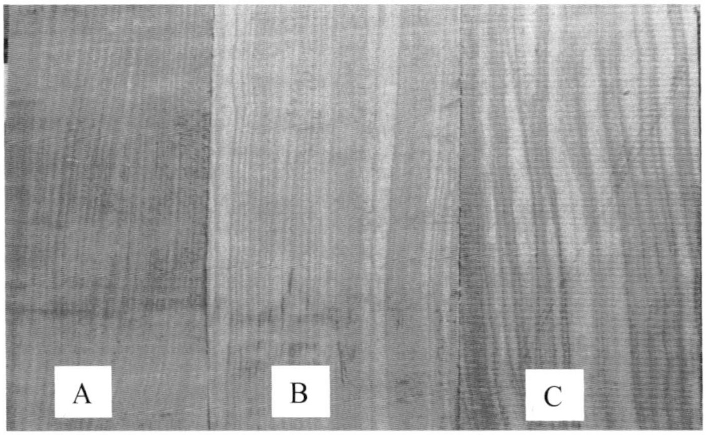

Figure 1. Classification of three grades of streak quality (class A, B and C, size 7 cm wide) (source of wood: PT. Saudara Indah Jaya Palu).

- Wood color index measurement was carried out on air dry heartwood powder with NF333 spectrocolorimeter. The system used is CIEL * a * b with 3 repetitions [9].
- Wood samples for measuring density and green/initial moisture content have sizes 2 x 2 x 2.5 cm (radial x tangential x longitudinal direction) using ISO standards [10]. The specific gravity was measured based on the oven dry weight and the oven dry volume. Procedure for measuring specific gravity and moisture content was based on SNI 03-6847 [11] and SNI 03-6850 [12].
- The influence of site on color index, initial moisture content and specific gravity were analyzed by independent t-test.

3. Result and discussion

Eboni has three wood zones i.e. sapwood, intermediate and heartwood. The intermediate zone is a transition zone from sapwood to heartwood and becomes a place for the synthesis of various chemical components to form a heartwood[13]. Intermediate wood has a thickness of 1-7 cm. Ebony BM has a thicker intermediate zone, while its sapwood is thinner than SL. The intermediate zone of ebony SL has the same thickness with ebony from Maros with 2-5 cm wide [14]. Intermediate zone is a formative stage of *D. celebica* which is not found in other Diospyros genus[15].

The colour of sapwood after being felled was cream to reddish brown, while the intermediate zone was reddish, and darker compared to sapwood. Ebony SL color was brighter than ebony BM (Figure 2). After drying in the room temperature, the color of both strike patterns were similar.

Eboni SL had a thickness of irregular black streaks between 1- 5 mm alternated with light brown streaks with a width of 4-15 mm. The number of black streaks was 17 per 8 cm. Eboni BM had a thin black streaks of 1 mm with light brown streaks 2-4 mm alternate regularly. The number of black streaks reached 28 strips in 7 cm (Figure 1). Based on local industry standards, ebony BM classified as grade B-A, while ebony SL classified as grade C. The interval between the black streaks (width of brown streaks) of *D.celebica* from Maros was 2-5 mm [15]. The better streaks quality are produced in drier agro-climate zones regions such as Donggala and Parigi Moutong. This requires further research [5]. The ebony industry acquires ebony BM from Domsol, Donggala District and Kasimbar and Sausu,

Parigi Moutong District which have agroclimate zones E1 and E2 according to Oldeman and Syarifuddin Classification [16].

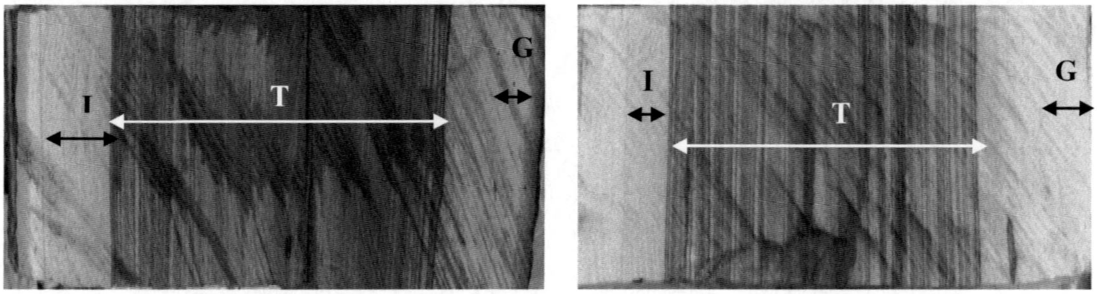

Batang Macis (A) Sarang Laba-Laba (B)

Remarks: G (Sapwood), I (intermediate), T (Heartwood)

Figure 2. Fresh cut color and streaks pattern of BM (A) and SL(B).

The color and streak pattern of heartwood is considered in processing of ebony. Darker heartwood, smooth and dense black streaks and thin brown streaks (ebony BM) are preferred as raw materials for furniture including butsudan, while ebony which has a brighter, wider black streaks and brown streaks (ebony SL), suitable for crafts and carvings. Ebony which has regular and smooth black streaks is preferred by consumers in Japan and China, while ebony which has thick and rare black streaks or wide brown streaks is preferred by consumers in Europe [7].

The results of color observations with CIELab system on air dried samples were shown in Table 1. The average brightness index (L *) is 14.07 ± 1.80, redness (a *) 7.59 ± 0.79, yellowish (b *) 11 , 75 ± 0.96, chromium (C) 13.99 ± 1.21 and hue (h) 13.95 ± 1.11. The t-test results (Table 2) showed that only the L * parameter was significantly different, while the other parameters were not significantly different between the both streak types. The color of the ebony SL wood was brighter, because the proportion of brown strip on the ebony SL was higher than the BM heartwood.

Table 1. Average of color index, moisture content and specific gravity.

Site	L*	a*	b*	C	H	Initial Moisture Content	Specific Grafity
BM	12.58	7.28	11.67	13.76	13.94	43.74	1.02
SL	15.57	7.91	11.83	14.23	13.96	46.11	0.99
Total	14.07	7.59	11.75	13.99	13.95	45.03	1.00

Table 2. Significance value for t-test (two-way) of color index, moisture content and specific gravity.

Source	L*	a*	b*	C	H	Initial Moisture Content	Specific Grafity
Site	0.042*	0.538	0.914	0.777	0.990	0.240	0.083

Remarks: Significance value <0,05, significantly different at α 0,05

The difference of wood color in *Acacia mangium* was related to extractive and phenolics contents. The lightness (L*) increased with decreasing phenolics contents. These means that darker color will be present in heartwood when higher phenolics contents is present [17]. Ebony BM has 12.96% and 13.15% of ethanol-toluene extraction and hot water extractive respectively, higher than ebony SL, 11.71% and 12.66% respectively [18]. In teak wood, the color index is more influenced by the site compared to wood age and climate. Smaller yellowish index (b *) (darker color) was associated with dry climate, deeper and more fertile soil [17, 19, 20].

The average of specific grafity of ebony was 1.00 ± 0.02. Specific grafity of ebony BM was 1.02 ± 0.02, was not significantly different but higher than ebony SL 0.99 ± 0.03 (Table 1). This specific gravity is slightly lower than previous research [21], which gained specific grafity 1.09 (1.01-1.27), but higher than ebony from Mangkutana, East Luwu, South Sulawesi and Donggala, Central Sulawesi 0.92 and 0.98 respectively [4,22].

The site had no significant effect on the specific gravity of ebony (Table 2), however, the specific grafity of ebony BM was slightly higher than ebony SL. The same results were found in teak wood (*Tectona grandis* L.) [23, 24,25]. Specific gravity is thought to be affected by the intensity of rainfall in site. *Eucalyptus globulus* and coniferous wood that grown in higher rainfall region have lower specific gravity [26,27]. Ebony SL had lower in specific grafity because grown in the wetter agro-climate zone (D1) compared to ebony BM (E1) [16].

Specific gravity value was affected by the thickness of the cell wall, the position in the stem and the proportion of wood cell. Wood with thicker cell walls has a higher specific gravity [28,29]. Further research in ebony wood is needed to prove this correlation.

The average of initial moisture content of ebony was 45.03 ± 5.10%. This moisture content value was lower than ebony from Maros of 49.2% [14]. The site did not affect the initial moisture content. The moisture content of ebony SL was higher than ebony BM. This is related to the specific gravity of ebony SL is lower than ebony BM. Specific gravity is the major determinant of maximum moisture content. Increase of specific gravity must decrease the volume of the lumina because the specific gravity of wood cell walls is constant among species. This decreases the maximum moisture content because less room is available for free water [30].

Ebony companies in Makassar and Palu have obtained ebony BM from Donggala and Parigi Moutong Regencies. Therefore,cultivation of ebony to produce the best streaks should be prioritized in these areas. Nevertheless,further research is needed regarding the characteristics of ebony sites in wider areas that produce BM and SL streaks. This information will be useful in the cultivation of ebony to produce streaks as desired.

4. Conclusion

This study determined the relationship between site factor and wood properties of ebony. The results showed that differences in streaks quality and brightness of heartwood color of the ebony grown in different sites were significant ($p<0,05$). Ebony from Parigi Moutong with BM streak had a grade B-A, better than that ebony from Poso. Ebony BM had a darker color, higher specific gravity and lower initial moisture content.

References

[1] Whitmore TC, IGM Tantra and U Sutisna 1989 Tree flora of Indonesia Check List for Sulawesi (Bogor: Forest Research and Development Centre)
[2] Badan Standardisasi Nasional 1990 SNI 01-2028-1990 Kayu Eboni Olahan
[3] Asdar M 2001 Struktur Anatomi Kayu Eboni (Diospyros celebica Bakh.) dari Kab. Luwu Utara, Sulawesi Selatan *Bull. Penelitian Kehutanan Balai Penelitian Kehutanan Ujung Pandang***7(1)** 1
[4] Asdar 2005 Karakteristik Anatomi dan Fisik Kayu Eboni (*Diospyros celebica* Bakh.) Dari Donggala Sulawesi Tengah Prosiding Seminar Hasil Penelitian (Bogor: Puslit Sosek dan Kebijakan Kehutanan)
[5] Asdar M, TA Prayitno, G Lukmandaru and E Faridah 2015 Sebaran, Potensi dan Kualitas Kayu Eboni di Sulawesi J. Agroland 22 (2) 1
[6] Kiding Allo M 2007 Ornamen Kayu Eboni Berdasarkan Lokasi Tempat Tumbuh di Sulawesi. Prosiding Ekspose Hasil Penelitian (Bogor: Pusat Litbang Hutan dan Konservasi Alam)
[7] Soerianegara I 1967 Beberapa Keterangan Tentang Djenis-Djenis Pohon Eboni Indonesia Rimba Indonesia XII 2-3-4 29

[8] Achmad A 2002 Gagasan Konservasi Flora Endemik (Diospyros celebica Bakh.) di Kawasan Wallacea Sulawesi-Indonesia Berita Biologi 6 (2) 337
[9] Lukmandaru G, T Ashitani and K Takahashi 2009 Color and Chemical Charactersization of Partially Black-Streaked Heartwood in Teak (*Tectona grandis*) J of Forestry Research20(4) 377
[10] Prayitno TA 1995 Pengujian Sifat Fisika dan Mekanika Menurut ISO(Yogyakarta: Universitas Gadjah Mada)
[11] Badan Standardisasi Nasional 2002 SNI 03-6847Metode Pengujian Berat Jenis Kayu dan Bahan dari Kayu dengan Cara Pencelupan dalam Air
[12] Badan Standardisasi Nasional 2002a SNI 03-6850Metode Pengujian Pengukuran Kadar Air Kayu dan Bahan Berkayu
[13] Hillis WE 1999 Recent advances in phytochemistry volume 33 ed J T Romeo(New York: Springer Science + Business Media LLC) p 432
[14] Prawirohatmodjo S 1977 Properties and Utilization of Some Indonesian Secondary Species PhD Thesis Australian National University Australia
[15] Hillis WEand Soenardi 1994 Formation of Ebony and Streaked Wood IAWA Journal15 (4) 425
[16] Oldeman and DSyarifuddin 1977 An Agroclimatic Map of Sulawesi (Bogor: Contr. Centr. Res. Inst. Agric)
[17] Moya R, RS Fallas, PJ Bonilla and C Tenorio 2012Relationship Between Wood Color Parameters Measured by the CIELab System and Extractive and Phenol Content in Acacia mangium and Vochysia guatemalensis from Fast-Growth Plantations Molecules17 3639
[18] Asdar M, TA Prayitno, G Lukmandaru and E Faridah 2016 Sifat Kimia Kayu Eboni pada Perbedaan Pola Strip dan Arah Radial Jur. Ilmu dan Kayu Tropis 14 (2) 165
[19] Derkyi NSA, H Bailleres, G Chaix, MF Thevenon, A A Oteng-Amoako and S Adu-Bredu 2009 Colour Variation In Teak (Tectona grandis) Wood From Plantations Across The Ecological Zones Of Ghana Ghana J. Forestry2540
[20] Thulasidas PK, KM Bhat and T Okuyama 2006 Heartwood Colour Variation In Home Garden Teak (Tectona Grandis) From Wet And Dry Localities Of Kerala, India J of Tropical Forest Science18 (1) 51
[21] Martawijaya A, I Kartasujana, K Kadir and SA Prawira 1981 Atlas Kayu Indonesia Jilid I (Bogor: Balai Penelitian Hasil Hutan Badan Litbang Pertanian)
[22] Asdar 2001b Sifat Fisik Kayu Teras dan Gubal Kayu Eboni (Diospyros celebica Bakh.) dari Luwu Utara Sulawesi Selatan Prosiding Ekspose Hasil Penelitian (Ujung Pandang: Balai Penelitian Kehutanan Ujung Pandang)
[23] Rulliaty S and M Lempang 2004 Sifat Anatomi dan Fisis Kayu Jati dari Muna dan Kendari Selatan. Jur. Pen. Hasil Hutan22 (4) 231
[24] Hadjib N, M Muslich and G Sumarni 2006 Sifat Fisis dan Mekanis Kayu Jati Super dan Jati Lokal dari Beberapa Daerah Penanaman Jurnal Penelitian Hasil Hutan24 (4) 13
[25] Marsoem SN, VE Prasetyo, J Sulistyo, Sudaryono and G Lukmandaru 2014 Studi Mutu Kayu Jati di Hutan Rakyat Gunungkidul III Sifat Fisika Kayu Jurnal Ilmu Kehutanan 8 (2) 75
[26] Miranda P, MH Almeida and H Pereira 2001 Influence Of Provenance, Subspecies, and Site on Wood Density in Eucalyptus globulus Labill Wood and Fiber Science33(1) 9
[27] Wiemann MC and GB Williamson 2002 Geographic Variation in Wood Specific Gravity: Effects of Latitude, Temperature, and Precipitation Wood and Fiber Science34(1) 96
[28] Panshin AJ and C de Zeeuw 1980 Textbook of Wood Technology Fourth Edition (New York: Mc Graw-Hill Book. Co).
[29] Shmulsky R and PD Jones 2011 Forest Products and Wood Science, An Introduction 6th Edition(West Sussex: Wiley-Blackwell)
[30] Forest Products Laboratory 2010 Wood Handbook—Wood as an Engineering Material General Technical Report FPL-GTR-190(Madison WI: Forest Products Laboratory) p 508

IOP Conf. Series: Earth and Environmental Science **359** (2019) 012006 doi:10.1088/1755-1315/359/1/012006

Efficiency of timber harvesting from natural forest in Indonesia

Soenarno*, **Dulsalam and Sukadaryati**

Forest Products Research and Development Centre
Jl. Gunung Batu No. 5, Bogor 16610, Indonesia

*email: soenarno@yahoo.co.id

Abstract. The efficiency of timber harvesting and damage to residual stands is an aspect that needs serious attention because it has impacts on 1) increasing logs production, 2) saving natural forest resources, and 3) reducing the volume of the waste left in the forest has potential to create a risk of greenhouse gas emissions. This paper aims to provide information on the efficiency of timber harvesting and residual stand damage caused by harvesting activities in natural forests. The results showed that: 1) Efficiency of harvesting natural forest timber ranged from 75–86% with an average of 80.6% before applying the low impact (RIL) and varied from 82–95% with an average of 87.6% for the RIL techniques; 2). Damage to residual stands due to logging of non-RIL and RIL ranged from 27.76–40.71% with an average of 31.97% and 5–40% with an average of 22.20%, respectively; 3). The application of RIL techniques and the existence of sustainable forest management certification could improve the efficiency of timber utilization.

1. Introduction

Forest resources, especially natural forests, have ecological benefits, socio-cultural and economic functions, hence the forests need to be managed sustainably to ensure their function and maintain -sustainability. Timber harvesting has a very important role in forest management activities, in terms of costs, productive time and forest sustainability. Harvesting activities may cause harvesting waste and damage to the remaining trees which difficult to avoid. The amount of timber harvesting costs reaches 30–50% of the total cost of forest management, the productive time of 40–60% of the total productive time of forest management and damage to stands of an average of 27.96% [1]. It was also stated that the most serious impact of timber harvesting that ignores the appropriate manner causes a decline in forest productivity.

In 1970-1990 through forest concession activities (as known as the permit to collect timber forest products in natural forests /IUPHHK-HA), forestry sector became the second largest state foreign exchange earner after oil and gas products, with annual production quota (APQ) logs from natural forests reach 20–40 million m^3/year. The government, namely the Ministry of Forestry, determined APQ based on the exploitation factor number (EF) of 0.7. Technically EF numbers are an indicator of efficiency in timber harvesting, where the more efficient timber harvesting activities possess a greater value of EF numbers and the less wood waste, and vice versa. Timber harvesting defined as part of branch-free stems which should have been used but due to various reasons, they were left in the forest. The efficiency of wood is a comparison between wood that is actually utilized to that potential portion of branch-free stems that can be used.

Management of Indonesian natural forests has been getting better with the application of RIL (Reduced Impact logging) and low-carbon impacts (RIL-C). Improved management of natural forests is characterized by an increase in efficiency of forest harvesting and the number of business licenses to utilize timber forest products in natural forests (BLUTFP-NF) which are certified to manage sustainable production natural forest. The average efficiency of timber harvesting in 1993 was 75%, but then during 1995–2013, had a dramatic increase to 83% [2]. In fact, the efficiency of harvesting timber produced by BLUTFP-NF that obtains both the mandatory and voluntary sustainable forest management (SFM) certificates through the forest stewardship council scheme, reaches more than 90% [3]. The number of (BLUTFP-NF) holders is 277 units with a concession area of 20.89 million ha and APQ round wood 9.10 million m^3/year [4]. The numbers were less than that of 1992 when the start of forest concession rights activities reached a total of 567 units and a concession area of 60.48 million ha or compared to that of 2010, 304 units with concession area of 24.69 million ha, distributed in Kalimantan, Sumatera, Sulawesi, Maluku and Papua [5]. It was also stated that at present, the main producer of logs from natural forests in Indonesia relies only on these locations.

This study aims to obtain information on the efficiency of harvesting timber in Indonesia's natural forest. Increasing the efficiency of harvesting natural forest timber is very important for the continuity of natural production forest management.

2. Materials and methods

2.1 Time, location and materials for research

The research was carried out in Bogor in September in 2015. The materials and tools used were publications about the efficiency of forest harvesting, writing instruments, and computers.

2.2 Research procedure

2.3 The research was carried out by gathering secondary data with the details in the following:

- Collecting material in the form of results of research on the efficiency of forest harvesting;
- Reviewing the results of research in terms of aspects of timber utilization, wood waste and efficiency of timber harvesting.
- Discussing the results of the study of the efficiency of timber harvesting
- Drawing conclusions from the results of the efficiency of timber harvesting studies

2.4 Data analysis

The method of analysis used in this research is an explorative descriptive method. A descriptive exploratory method to get an idea of how the efficiency of harvesting is different for a certain period of time. The average analysis is used to see the direction of interrelation between time periods.

3. Results and discussion

3.1 Timber harvesting natural forest and reduced impact logging on natural forest

Timber harvesting is a series of forestry activities that convert trees and other biomass into forms that can be moved to other locations so that it is beneficial to the economic and cultural life of the community [5]. Initially, timber harvesting was carried out conventionally using a system called Indonesian selective logging (TPI) and since 1989, the system has been changed to Sylvicultural System of Indonesian Planting and Selective Cutting (SSIPSC). There was poor planning in the conventional timber harvesting, limited training for operators of the chainsaw and skidding tractors. Furthermore, there was no control both at the time of implementation and at the post-harvest operations. At that time, the orientation of harvesting natural forests was prioritized as the country's foreign exchange earner, thus ignoring the principle of sustainable forest management. As a result, it causes: (a) excessive open land; (b) unnecessary felling of trees; (c) soil damage and excessive erosion; and (f) logging waste happened not only at the felling site but also at Temporary Timber Collection Sites (landing places) because of the testing process and measurement of the logs (grading & scaling).

The timber harvesting process generally consists of several main activities, as follow [6]:

- Operating the stump (stump operation), the felling of trees and the establishment of the beginning of the log;
- Skidding, which is to move the logs as a whole or in the form of logs from the felling place to a landing place, with a distance of only a few hundred meters;
- Loading, which is raising wood onto the transport equipment on the landing places;
- Main transportation (major transportation), which is transportation the logs from the landing places to the final destination (Log yard or log pond);
- Unloading namely log unloading at the destination.

Recognizing the remaining limited production forests and the risk of damage to forests by conventional timber harvesting, the technique of harvesting timber in natural forests that are more environmentally friendly, namely Reduced Impact Logging (RIL) techniques began in 2001. The RIL technique initially emphasizes on improvement of harvesting planning, implementing the controlled timber harvesting and measures to prevent damage from post-harvesting operations [8]. In its implementation, in order to achieve the goal of sustainable forest management in the TPTI system, there are three criteria for the RIL concept that must be obeyed, which consist of [7]:

- Timber harvesting is not allowed to exceed forest site production capacity so that it does not threaten the sustainability of timber production in the next cutting cycle;
- Forest damage does not exceed the capacity of ecosystem recovery so that forest damage that occurs immediately recovers and forest growth (increment) is maximal;
- Timber harvesting may not cause a scarcity of commercial type.

The activity of making the notch and notch back in felling trees to determine the fall direction must be undergone correctly. Determining the correct direction of felling the tree are as follows [8]:

- Wherever possible to avoid places with many obstacles, such as rocks, fallen trees, stumps, ditches;
- If the tree is located on a slope or cliff, then the direction of felling is directed to the top of the slope or at least parallel to the contour;
- Working towards a place where the stand is relatively less;
- The direction of felling is pursued in accordance with the direction of skidding or which makes it easier and
- In areas with flat conditions the direction of felling the tree is adjusted to the canopy and position of the tree.

3.2 Development on the efficiency of the timber harvesting in forest production

3.2.1 Periode before RIL system. It is estimated that the amount of wood harvesting waste ranges from 25-35% or reaches ± 20 million m^3 [9] [10] [11]. The amount of wood waste did not include waste in the form of damaged trees, branches, twigs, stumps and waste as a result of the production process. Meanwhile, the results of a study of the Forest Products Research Institute in 1976–1978 stated that the amount of waste that occurred in 24 forest companies ranged between 25–48% or reached 5–12.6 million m^3. This means that the timber harvesting activity is still wasteful and the harvesting efficiency is low, which ranges from 52–75%. Some research results suggest that the efficiency of harvesting wood for the use of stems to branch free stems is presented in Table 1 [2].

Table 1 Efficiency of harvesting timber from a dry land natural forests

Year	Harvesting effect (%)	Wood waste (%)
1985	80	20
1988	83	17
1993	75	25
1995	86	14
1996	80	20
1997	80	20
Average	80.6	19.4

Source : [3]. Data is purposely processed

The efficiency of harvesting timber which is calculated to a diameter of 30 cm is still very low, ranging from 42–56%. This means that wood waste left in the forest reaches a range of 44–58%. Meanwhile, the results of research on the damage to residual stands in 1974-1993 ranged from 23.00 to 40.71% with an average of 32.20% or included in the category of damage to medium-level stands (Table 2) [12] [7].

Table 2 Average damage to standing trees in conventional harvesting techniques

Researcher	Location	Average damage of residual stand (%)
Tinal & Pallenewn (1974)[a]	East Kalimantan	36.40
Muhandis (1976)[a]	Puru Island	23.00
Firdinandus (1978)[a]	West Kalimantan	40.71
Elias (2016)	Borneo	27.96
	Average	32.2

Remark: a = [7].

3.2.2. The Period after RIL system. From 2001 until now natural forest management by business license for utilization of timber forest products of natural forest has changed even better with the implementation of the principles of Sustainable Production Forest Management (SPFM) and reduced impact logging (RIL) harvesting techniques. Selection of the system and harvesting techniques must meet the requirements to- 1) minimize stand and soil damages, 2) reduce timber harvesting waste, and 3) provide good quality timber production [13].

Management of natural forest production is getting better as indicated with more efficient timber harvesting and lower damage to residual stands. Increasing efficiency of timber harvesting means increasing the volume of round wood production and ensuring more sustainability. This is simply due to the increase of production volume obtained without increasing logging area and the number of the cut trees. The Ministry of Environment and Forestry perceived direct economic benefits in the form of increase non-tax state revenues.

Advantages of the RIL timber harvesting are as follows [14] : 1) damage to fewer stands (41%); 2) the area covered by the skidding track is relatively the same (50%); (3) areas damaged by fewer road constructions (40%); (4) lower canopy opening (33%); (5) the volume of logs lost as waste was less (33%). Nevertheless, the fact in the field some IUPHHK-HK still apply RIL technology partially. This has led to the efficiency of optimal timber harvesting and many kinds of woods logging waste in the forest. The results of the 2013-2017 study had conducted in several IUPHHK-HA showed that the efficiency of timber harvesting averaged 86.3% (Table 2) and left wood waste on average by 13.7% or still far from the target set by RIL which is a maximum of 5% [15]. Therefore, the active role of the Agency for Development, Research and Innovation are still needed to continue to innovate zero-waste and environmentally friendly timber harvesting methods. The efficiency of timber harvesting at several BLUTFP-NF in 2015–2017.

Table 3. Efficiency of timber harvesting at several BLUTFP-NF in 2015–2017

IUPHHK-HA/Location	Average of (%)	
	Efficiency of timber harvesting	Damage to residual stand
PT. Inhutani I Sambarata (East Kalimantan, mandatory, 2015)	86	32.42
PT Segara Indochem (East Kalimantan , mandatory, 2015)	77	No data
PT Karya Lestari (East Kalimantan, mandatory, 2015)	85	No data
PT.WIjaya Sentosa (Papua, voluntary, 2016)	88	25,18
PT. Jati Dharma Indah Plywood Industries (Papua, mandatory, 2016)	86	20.95
PT. Prosperity of Berkah Timber (East Kalimantan, voluntary, 2014)	92	No data
PT. Kayu Tribuana Rama (Central Kalimantan, voluntary, 2016)	84	24.93
PT.Ratah Timber (East Kalimantan, voluntary, 2014)	88	10.43
PT.Roda Mas Timber (East Kalimantan , voluntary , 2014)	91	5.94
PT. Belayan River Timber (East Kalimantan, voluntary,2015)	86	35,55
Average	86.3	22.2

Sources: [15] (Data purposely processed)

Based on the research results in Table 1 and Table 3, it can be illustrated that the application of RIL technology can reduce wood harvesting waste, as in Figure 1.

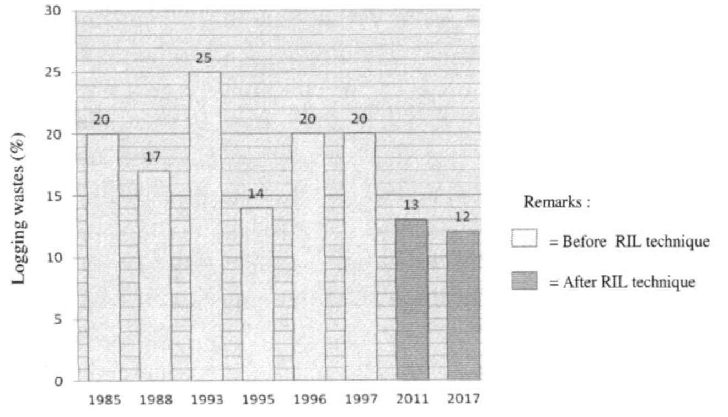

Figure 1. Dynamical changes of potential logging wastes

When the average of national log annual production quota is 9.1 million m^3/year [4] it is estimated that the potency of logging wastes reaches ± 1.128 million m^3/year. The potency of logging waste has not been utilized yet by BLUTFP-NF because of economic considerations. The research results in

IOP Conf. Series: Earth and Environmental Science **359** (2019) 012006 doi:10.1088/1755-1315/359/1/012006

2015 on BLUTFP-NF indicated that the logging efficiency depends on felled tree diameter [15]. The relationship between wood utilization efficiency shown by the value of the exploitation factor and tree diameter is presented in Figure 2.

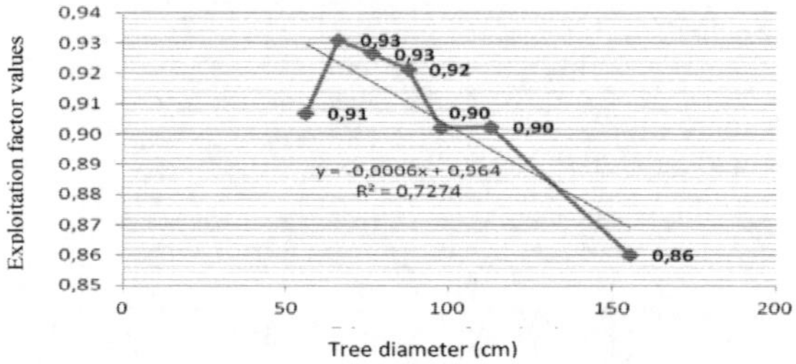

Figure 2. Relationship between felled tree diameter and exploitation factor

Technically, it's easy to understand because cutting large-diameter trees is more difficult than smaller diameter one. Large trees in natural forests have generally buttresses, causing an imperfection for making notches. As a result, there is often an error in the direction of felling the tree causing wood to break. The broken wood not only occurs at the base of the stem but also at the end of the tree trunk even sometimes in the middle of the tree. Many factors that influence the efficiency of timber harvesting are technical, economical and ecological as well as policy aspects [2]. Of these four factors, technical logging and management policies have a greater influence on the efficiency of timber harvesting. The facts in the field indicate that differences in management conditions of BLUTFP-NF more influential than other technical factors. The size of the role of each factor varies between BLUTFP-NF and each other. This is in line with another research[19] that the efficiency of timber harvesting is mainly influenced by the systems and techniques of timber harvesting activities. Timber harvesting techniques cannot be separated from the component of timber harvesting activities, such as determining the direction of felling trees, tree cutting, skidding, loading and unloading as well as transportation. One of the most principle influences on the efficiency of wood harvesters is the habit of cutting methods carried out by chainsaw operators and skidding tractor operators who work on the basis of practicality to obtain a volume of prime quality timber. [16]

3.2.3. Impact of timber harvesting efficiency. The increasing efficiency of timber harvesting not only benefits the holders of business license for timber forest products utilization of natural forests but also the Ministry of Environment and Forestry. The impact for the holders of business license for timber forest products utilization of natural forests is that the holders will an increase in production volume which has implications for revenue from the sale of logs. Besides that, the government namely the Ministry of Forestry will increase non-tax state revenue and reduce the rate of forest exploitation. As an illustration, with the assumption that annual production quota of 9.1 million m3 / year, the selling price of logs is 2.2 million/m^3, non-tax state revenue is in accordance with the regulations (Rp 271.750/m^3), basic price of logs of Rp 760.000/m^3 and the potency for timber to be harvested ranges from 20-30 m^3/ha, the impact of increasing the efficiency of timber harvesting is presented in Table 4.

Table 4 Impacts of the increasing timber harvesting efficiency

Discription	Timber harvesting efficiency (%)		
	Base line determination of JPT	Before RIL	After RIL
Annual production quota (million m³/year)	9,10	10, 35 4	10.58
Additional wood volume (million m³/year)	-	0.964	1.48 0
Additional value of timber sell (x IDR trillion/year)		2 , 120	3 , 256
Additional acquisition of PNBP (x IDR billion)		261,967	4 02, 190
Extensive felling savings (ha/year)		32,133 - 47 .700	49,000 - 74,000

Table 4 above shows that the increased efficiency of timber harvesting increased national log production volumes ranging from 0.964 to 1.48 million m^3/year with a value of Rp 2.120–Rp 3.256 trillion/year. If it is associated with a deficit of raw material requirements for the national wood processing industry, an average of 23.3 million m^3/year, the increasing efficiency of timber harvesting can contribute 4–6%. In addition, the increasing efficiency of timber harvesting can increase the potential of non- tax state revenue by Rp. 288,055–Rp. 402,190 billion/year or 7–9.8% of the average non- tax state revenue target during 2011–2015 amounting to Rp. 4.1 trillion/year [4]. From Table 3 it can be seen that increasing efficiency also has an impact on the reduced area of timber harvesting which ranges from 30,000–74,000 ha/year.

4. Conclusion and suggestions

Natural forest management in Indonesia has improved, characterized by several improvements of efficiency in timber harvesting of 10.6–16.3%, additional wood volume of 0.954–1.480 million m^3/year, increase in non tax state revenue of 7–9%, reduction in felling area of 32,133–74,000 ha / year and increased fulfillment of wood deficits of 4–6% per year. Increased efficiency of timber harvesting has an impact on a) increasing round wood production, b) reducing wood waste in the forest, c) savings in natural forest resources and d) increasing non-tax state revenue. Better studies and innovations in the method of harvesting natural forest timber are still needed for the innovation of zero waste and environmentally friendly.

References

[1] Elias, Applegate, G., Kartawinata, K., Machfudh, & Klassen, A. (2001). *Guidelines for Indonesia's reduced impact logging* (Bogor, Center for International Forestry Research)
[2] Idris, MM, Dulsalam, Soenarno & Sukanda. 2013. *Revision of exploitation factors for logging optimization*. Proceedings of the Exposition of the 2012 Research Results on Forest Engineering and Forest Products Processing (Bogor, The Center for Research and Development on Forest Engineering and Forst Products Processing)
[3] Soenarno, Dulsalam, & Endom, W. 2013. Factors of exploitation in limited production forests in IUPHHK-HA PT Berkah Kemakmuran Timber . *Journal of Forest Product Research* (Vol. 31).
[4] Ministry of Environment and Forestry. 2016. *Ministry of Environment and Forestry Statistics 2015* . Jakarta: Data and Information Center.
[5] Suparto, R.S. 1982. *Modern Forest Exploitation Book* (Bogor, Faculty of Forestry IPB).
[6] Wiradinata, S . 1989. *Logging planning for decreasing the logging wastes and the residual stand damages*. Proceedings of the Seminar on Forest Exploitation (Bogor, Forest Products Research Institute)

[7] Elias. 2008. *Forest Area Opening*. IPB Press. Bogor.
[8] Elias. (2016). *Application of reduced impact logging in order to reform forest exploitation and corruption in Indonesia's tropical natural forest management* (Bogor, Faculty of Forestry, IPB)
[9] Klassen, A. (2011). *Consideration in planning low impact logging* . (Hasbillah, Ed.) Jakarta: Tropical Forest Foundation.
[10] Silitonga, Jam & Hiras S. (1980). *Supervision is needed to reduce waste and ensure forest sustainability*. Proceedings of the Seminar on Forest Exploitation in Cisarua on July 8, 1980 . (Bogor, Forest Products Research Institute)
[11] Sumitro, A. (1980). *Skidding methods to reduce waste and damage to stands live in forests outside Java.* Proceedings of the Seminar on Forest Exploitation in Cisarua on July 8, 1980 (Bogor, Forest Products Research Institute)
[12] Wiradinata, S & Widarmana, S (1980). *Logging planning for reduced waste and damage to residual stands*. Proceedings of the forest exploitation seminar (Bogor, Forest Products Research Institute)
[13] Wahyudi. (2013). *Silviculture system in Indonesia theory and implementation* . Palangka Raya: Palangka Raya University .
[14] Ruslandi. (2013). *Application of low-carbon impact logging (RIL-C)* . Jakarta: The Nature Concervancy.
[15] Soenarno, Endom, W., Basari, Z., Suhartana, S., Dulsalam, & Yuniawati. (2016). Factors of forest exploitation in the East Kalimantan Sub Region. *Journal of Forest Product Research* , *34* (4), 335–348.
[16] Mansyur, A., Tirkaamiana, M.., & Sutejo, H. (2013). Harvesting waste and IUPHHK-HA exploitation factors PT Rizki Kacida Reana, Paser Regency, East Kalimantan Province. *AGRIFOR* , *XII* , 116–131.
[17] Putz, FE, Dykstra, DP, & Heinrich, R. (2000). *Why Poor Logging Practices Persist in the Tropics* (Vol. 14).

IOP Conf. Series: Earth and Environmental Science **359** (2019) 012007 doi:10.1088/1755-1315/359/1/012007

The Economic and Environmentally Friendly Tree Felling Techniques in Natural Forest

S Suhartana* and Yuniawati

Forest Products Research and Development Centre,
Jl. Gunung Batu No. 5, Bogor 16610, Indonesia

*email : sona.suhartana@gmail.com

Abstract. Improper tree felling techniques may reduce productivity and log supply. This leads to an increase in production cost. Timber Utilization Efficiency (TUE) in natural forests has not been optimally implemented, thus leaving residual stands with tremendous damages. This study was carried out in late 2015 in PT Inhutani II, North Kalimantan to investigate the implementation of felling techniques and to analyze felling productivity, TUE, production cost, and residual stands damage. Productivity, felling cost, and TUE data were tabulated and averaged. Results revealed that: (1) tree felling techniques implemented on the site were the reduced impact logging (RIL) technique; (2) the average felling productivity and production cost were 38.308 m^3/hour and IDR 13,935.10/m^3, respectively; (3) the average TUE was 92%, with only 8% generated-wastes; and (4) the average destruction of residual stands was 13.54%, which comprised of canopy damage, broken/cracked, stem injury, and leaning.

1. Introduction

Tree felling or tree logging presents an initial step in realizing the production of timber from the inside forest. The felling in natural forests uses chainsaws numerously that comprise of several types and varying machine power. The logging process cannot be separated from productivity and production costs aspects.

[1] alleged that there were several factors that could affect how large or small the cost and productivity would be at any activities, and also could render the harvest to be more efficient and effective, among others: (1) Moving duration for chainsawmen, roads and undergrowths should not be necessarily cleaned to shorten working duration for the chainsawmen in their movement from one tree to another; (2) Establishing the cutting steps; (3) Tree diameter, it can affect the productivity as the smaller the diameter then the shorter would be the cutting duration or the faster the cutting work; and (4) kind and specification of tools, a chainsaw with high rotating speed (rpm) could accelerate the felling work (or shorten wood-cutting duration) to fell down one tree.

Efficiency closely relates to the logging wastes, which are simply the residuals that consist mostly of cut-wood pieces abandoned in the exploited forests. These wastes are merely organic stuff- which does not pose any harmful threats to the environments, but their enormous amounts could indicate inefficiency degree in the forest logging. Tree stems are not entirely removed from forests, but some part of them is uncontrolled stay inside as wood wastes. Wood wastes of logging, according to [2] are defined as wood stuff that was not or still not utilized that resulted from the forest logging. Those wastes could come from allowable cut trees that composed of residues from stem division (trimming), stumps, twigs, and end portion, whereby their generation could occur on the felling sites, loading point (TPn), and log yard (TPK).

Logging efficiency is mostly affected by the systems and techniques for such logging activities. Logging techniques should not neglect the stages in the tree felling that present the essential components in the logging activities; and should not either disregard activities in determining the tree-felling direction, tree-felling implementation, bucking, log skidding, log-bark peeling, and hauling. Conducted properly felling techniques usually owe the precise direction in tree-felling downing, as low as possible, could supposedly reduce or minimize the degree of timber damages at the felling sites.

[3] asserted that activities of logging at natural forests should implement the techniques in such a way able to reduce the residual stands damage in maintaining their potency to secure the sustainability of natural forests. Activities in logging at natural forests which are indicatively potential to cause the residual stands damage are in tree felling activities and log skidding. Techniques of tree felling which are performed carefully in determining the felling direction and preparing the cutting steps are supposedly able to reduce the residual stands damage.

Residual stands refer to the stands after their tree origins have been selectively felled down. Those stands could become essential assets for further endeavors, which consist of nurtured trees and accompanying trees [4]. Nurtured trees pertain to the trees that should be taken care after being selectively cut, which consist mainly of young and healthy commercial trees. Meanwhile, the accompanying trees refer to the trees that make up other stands than nurtured trees. The degree of residual stands damage in natural forests could be affected by the techniques which are implemented for logging.

According to [5] the degree of residual tree stands damage was figured-out based on the ratio between the number of trees which were damaged due to the activities of timber logging and the number of trees that already existed in the forest area subtracted by the number of trees which were harvested. Residual stands that must be available according to the regulation endorsed by TPTI (Tebang Pilih Tanam Indonesia / Selective Tree Cutting for Indonesia's Planting) should in number correspond to 2000 measuring plots/ha. Forty percent (40%) out of those 2000 plots or be numbering to 800 plots must contain "seedlings"; or in other way, 1000 plots (out of 2000 plots) are available with minimal 400 plots (of 1000 plots) spared, and 60% of them (or 240 plots) contains stakes. As such, minimally 100 plots (out of 400 plots) should also be available, whereby 75% of them or 75 plots contain poles and the remaining (25 plots) contains entirely the core trees (100%).

Further, [6] alleged that forest governance in Indonesia should abide strictly by the sustainability principles that attempt earnestly to obtain the results which were maximally kept. Such sustainability principles desire to get the results continuously and if possible increasing either qualitatively or quantitatively. Meanwhile, maximal principles serve as a base which is adopted to utilize forest resources optimally. Sustainability principles can be assured should the volume of timber in the felled tree in particular duration can be more or less equal to the wood volume which grows in the same duration.

Contemplating those backgrounds and other previously narrated aspects, this scientific writing intends to look into and assess the tree-felling techniques as adopted, felling productivity, costs of felling production, and the residual stands damage in natural forests.

2. Materials and methods

2.1 Time, location and materials for research

The research was conducted at a forest concession administered by PT Inhutani II (the State Forest Company in Indonesia), which belonged to the working site of Sei Tubu's Unit, Malinau Regency, Province of North Kalimantan. The research was carried out in late 2015. Materials used were paints, paint brushes, and plastic ropes. Equipment employed in this study comprises of chainsaws and meter-measuring devices, writing utensils, and computers.

2.2 Research procedure

The research was carried out through collecting primary and secondary data with details in the following:

2.2.1. Determining the location for research in order to represent the physical condition of local environments;

2.2.2. To prepare 10 research plots that entirely occupied one-hectare area, which each measured approximately 100 m by 100 m (Figure 1)

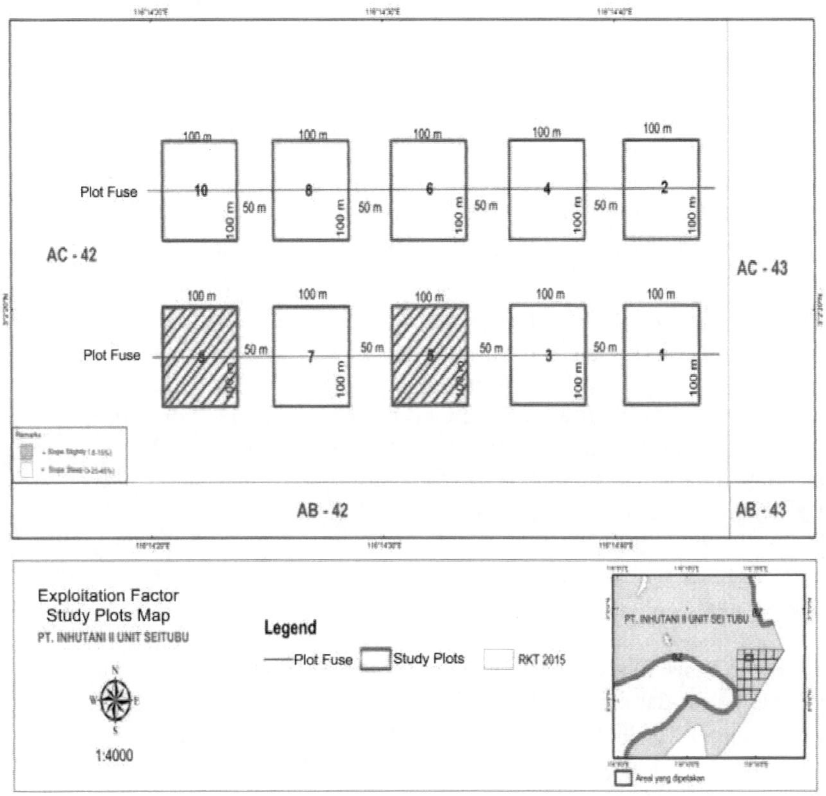

Figure 1. Study Plots map

2.2.3. The first research plot was determined purposively; and the further plots (second, third, fourth, etc) were determined later successively and systematically, whereby the spacing distance between plots was equal to 50 m;

2.2.4. At each of the research, plots were conducted inventories on the tree stands with their diameter equal to 20 cm or greater; and then diameter at breast height (dbh) and the height of trees vertically up to the point free of branches (clear bole) were measured.

2.2.5. Conducting the tree felling

2.2.6. The practical scheme regarding wastes and timber that could be utilized is presented in Figure 2

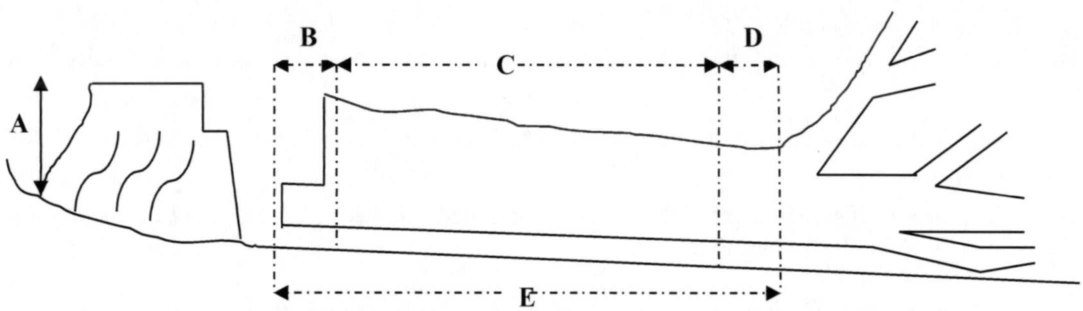

Remarks: A = Stump wastes; B = Lower-end waste; C = Usable portion of wood; D = Upper-end waste; E = Clear bole

Figure 2 Size measurements on usable wood-portions and tree felling wastes.

2.3. Primary data.

Measuring the diameter of wood stem portion (round shape), and its length starting from the lower cut-portion all the way until the first branch-free portion (Figure 2); Conducting bucking, using chainsaw; measuring the size of the wastes, generated from the tree felling. In this study, logging wastes referred to the portion of stems or portion of trees that were allowed to be cut, but not utilized and then just laid in the forest land. Those wastes in shape could consist of stumps and branch-free stems (due to their shape, size, and objectionable conditions, unable to be utilized). In general, the portion of trees is composed of two groups, namely stumps and branch-free stems.

Stumps pertain to the lower portions of trees with their position beneath the so-called felling cut and back cut. The stump dimension as measured comprises of diameter and height. In this research, the procedures for measuring the allowed stump height were referred to the RIL's technical guidelines [7]. Clear bole refers to the main stems starting from the buttress up to the first branch-free portion. Wastes from the clear bole can consist of short-cut wood pieces or logs. Short-cut pieces imply the part of main stems that contain defect or damages and therefore need to be cut. Short-cut pieces can also include twisted defects, crocodile-shaped knots, decays (rots), holes, and breakages.

Recording the damages that occurred to the residual stands. The definition of tree is the forest plants that exhibit their diameter equal to or greater than 20 cm. To estimate the tree damages, it uses the criteria enacted by [4], which mentioned: Tree canopy damages physically reach greater than (>) 30%, or there occur broken branches (or broken big branches); Stem injury occurs greater than (>) one fourth (1/4) of the stem circumference, with its length ≥ 1.5 m; Stem injury occurs greater than (>) one fourth (1/4) of the stem circumference, with its lateral length ≥ 0.15 m; The cut root or one third (1/3) of the buttress is physically damaged.

Relevantly, the trees are regarded as damaged, if it could abide by one or more of those three criteria (a, b, c) stipulated as above. Measurement of productivity, production cost, TUE, is as follows: a). Felling productivity is calculated by recording the duration for tree- felling that uses the so-called null-stop method, and figuring-out the volume of timber from the felled trees; b). The production cost of tree felling is by recording all the incurred expenses such as fuel, lubricating oils, wages, productivity, and depreciation costs, maintenance/repairment costs, interests, insurances, taxes, and salary costs; and c). TUE is obtained by recording diameter of the stem at the lower end portion, diameter of the stem at the upper-end portion, tree height, stem length, and stump height, together with other supporting data.

IOP Conf. Series: Earth and Environmental Science **359** (2019) 012007 doi:10.1088/1755-1315/359/1/012007

2.4. Secondary data.

Secondary data were obtained from literature and by recording the plan and realization of annual production for five-year duration of the company.

2.5. Data analysis

Data analysis was conducted by arranging them in tabulation form. Parameter for the analysis as implemented was the average values (means). Processing results led to the acquirement of quantitative information about felling productivity, the production cost of the felling, TUE, and residual stands damage.

2.5.1. Timber volume (Formula 1):

$$V = 1/4 \pi D^2 \times L \qquad \text{.......(1)}$$

where: V = timber volume (m^3); D = average timber diameter (m) (by averaging-out the diameter at upper portion of the stem and diameter at lower portion of the stem; π = phi constant (22/7); and L = length of stem (m).

2.5.2. Felling productivity (Formula 2):

$$Pt = V / t \qquad \text{.......(2)}$$

where, Pt = felling productivity (m^3/hour); V = volume of timber, (m^3); t = felling duration (hours)

2.5.3. Production cost of the tree felling (Formula 3):

$$Bt = (Bm/Pt) + U \qquad \text{.......(3)}$$

where: Bm = machine-operation cost (IDR / hour; IDR = Indonesian currency), calculated using the formula adopted by [8]; Pt = tree felling productivity (m^3/hour); U = wages (Rp/m^3).

2.5.4. Timber utilization efficiency (TUE) (Formula 4):

$$Ef = (Vp/Vm) \times 100\% \qquad \text{.......(4)}$$

where: Ef = efficiency of utilization (%); Vp = volume of timber as acquired (logged) (m^3); Vm = volume of timber that should have been utilized (m^3)

2.5.5. The degree of residual stands damage (Formula 5) as adopted from [9]:

$$KP = \frac{RP}{P - Pt} \times 100\% \qquad \text{.......(5)}$$

where: KP = degree of residual stands damage (%); RP = number of trees which were damaged (trees/ha); P = number of trees with diameter equal to or greater than 20 cm, before being felled ; Pt = number of trees felled.

3. Results and Discussion

3.1 Felling productivity

Activities of tree felling signified as the initial stage in timber harvest with the aim to obtain raw material to supply wood-processing industries in adequate amount and with qualities satisfying the desired requirement. Table 1 depicts that average felling productivity afforded by PT Inhutani II reached 38.303 m^3 (of the corresponding timber) with average volume of the felled trees equal to 6.735 m^3 in 10.5-minute duration.

Average productivity in this research was greater than that performed by [10], whereby the average felling productivity in the natural forest that used Stihl *chainsaw* reached 26.1 m^3/hour, with average felling duration equal to 4.57 minutes per tree, and average volume of the felled trees as much as 1.900 m^3 (of equivalent timber). Research results by [11] conducted in Turkey's natural forest brought out the productivity of tree felling at 19.19 m^3/hour, with average felling duration 5.69 minutes and average felling volume equal to 1.82 m^3.

Although the average felling productivity in this research was greater than those performed by [10] and [11] the average felling duration in this study, however, was also longer. Such duration could happen as the topography condition of the research area belonged to the category as predominantly steep, thereby causing the machine operators to take a longer time to move from one tree to another. Also, the area condition brought about additional difficulty to the operators in determining the tree-felling direction. Besides those factor hindrances, the skills afforded by operators should be better in the understanding (learning) stage. In addition, the health and safety of the operators should be taken as a priority. Moreover, such high productivity of felling in this research owed to the obtained volume of timber (from the tree felling) which was greater than from the two other research results.

The research area which was dominated by steep topography could affect the average timber productivity from the tree felling. [12] asserted that activity of causing inconvenience, exhaustion, and unsafe feeling due to working under the worry condition that occurred almost frequently during their work. Accordingly, this could end up with the decreasing felling-productivity.

Several factors that could affect the productivity of felling in the natural forest were among others: 1) Determining the felling direction: it would be better before the felling work to determine or approximate the direction of the trees to be felled down. If the felling direction were right, proper, and precise, then it could minimize the working-accident risk and also reduce the residual stands damage; 2) Preparing the felling-cut: as such there was requirement regarding the depth in preparing the felling-cut which consisted of two main items, consecutively cut base and cut roof. The cut base should be prepared earlier with the depth about 1/5 – 1/3 of tree diameter (at dbh). Afterwards, the cut roof could be made at angle 45° to the cut base. In this way, it brought out pieces which were so-called cut mouth. If the preparation of felling-cut did not follow the regulation, then breakage could occur to the lower end of the tree stems, thereby generating a higher amount of wastes. Besides, the felling-cut which was made too deep could cause the trees to fall down prematurely before their intended time; and there would also occur the so-called barber chair in that wood fibers would project or protrude out above the stumps due to a mistake in preparing the felling-cut; and 3) Preparing the back cut: the height of back cut was approximately 1/10 of the tree diameter beginning from the line along the base cut. The back cut was made by cutting the tree horizontally at the height above up to the so-called hinge wood. The function of hinge wood was as a drive in controlling and orienting the direction of the felled trees [12].

Table 1 Average of felling productivity at PT Inhutani II *).

Sampling plot for the research	Tree No	Timber volume of felled trees (m^3)	Felling duration, (minutes)	Timber productivity from the tree felling, (m^3/hour)
I	1	9.464	15.0	37.855
	2	11.462	17.0	40.455
	3	9.125	14.0	39.106
II	4	7.078	11.8	35.991
	5	7.063	11.7	36.220
	6	5.444	8.0	40.833
	7	5.815	8.2	42.547
	8	3.956	6.8	34.909
	9	7.219	11.9	36.398
	10	4.171	7.0	35.752
III	11	2.703	4.0	40.539
	12	4.008	6.2	38.789
	13	4.963	7.5	39.704
	14	9.625	15.1	38.247
	15	12.140	18.1	40.244
	16	5.440	8.0	40.804
	17	4.338	6.0	43.379
IV	18	9.585	15.2	37.836
	19	7.544	10.5	43.108
	20	4.447	6.6	40.423
	21	11.965	17.5	41.022
	22	4.753	7.4	38.541
	23	2.508	4.1	36.705
	24	4.780	7.3	39.289
V	25	6.243	9.1	41.160
	26	4.667	6.5	43.078
	27	6.325	10.0	37.949
	28	15.116	21.1	42.984
	29	7.530	11.1	40.702
	30	3.492	6.2	33.797
VI	31	11.588	18.1	38.414
	32	2.798	4.4	38.154
	33	7.424	11.5	38.735
	34	3.961	6.3	37.725
	35	7.478	11.6	38.677
	36	3.832	6.5	35.375
	37	2.453	4.3	34.230
VII	38	7.823	11.1	42.286
	39	3.003	5.5	32.759
	40	6.251	10.0	37.505
	41	9.975	15.2	39.376
	42	10.029	17.1	35.189
	43	3.870	6.0	38.699

VIII	44	4.089	7.6	32.285
	45	6.878	10.1	40.859
	46	5.777	8.5	40.776
	47	3.609	6.3	34.367
	48	3.752	6.0	37.518
	49	7.148	12.0	35.739
	50	10.401	17.5	35.660
IX	51	3.453	6.0	34.534
	52	13.680	20.0	41.039
	53	9.182	15.3	36.007
	54	9.530	15.1	37.866
	55	9.327	15.5	36.104
	56	10.457	17.1	36.690
X	57	4.749	7.6	37.489
	58	3.571	5.2	41.200
	59	7.849	12.1	38.918
	60	7.189	12.0	35.945
Sum		404.093	630.4	2298.487
Average		6.735	10.5	38.308

Note : *) The State Forestry Company in Indonesia

3.2 Cost of tree felling production

Average of timber productivity from the tree felling as obtained (Table 1) could be used as a divisor in the calculation of average cost for felling production. The resulting calculation regarding the cost components in tree felling and average production cost also in the tree felling is presented in Table 2. The tree-felling cost per m^3 of timber could be calculated through the ownership cost and tool-operation cost as follows: (1) Price of Stihl 070's chainsaw tool was Rp. 15,950,000/unit; (2) service life of the tool lasted for one year or 1,000 hours; (3) Insurance equal to 3%/year; (4) Bank interest as much as 12%/year; (5) Taxes as high as 2%/year; (6) price of gasoline worth IDR 6.500/liter; (7) salary for operators and assistant-operators was IDR 13,000/m^3; (8) Working duration per day reached 8 hours; and (9) Tool power equal to 6.5 HP. From those cost data and using [8] formula, then the cost components could be calculated as presented in Table 2. Further, the values for each of the tree-felling production costs were obtained by dividing the total cost for enterprise (endeavor) by each of the felling productivities, as depicted in Table 3.

Table 2 Cost components for the tree-felling operation using chainsaw machine.

Cost components	Amount (IDR/hour)
Depreciation expenses)	14,355
Insurance expenses	287.1
Interest expenses	1,435.5
Tax expenses	191.4
Fuel expenses	4,563
Oil expenses	456.3
maintenance expense	14,355
Machine expenses	35,643.3
Wages expenses (IDR/m^3)	13,000

Table 3 Average of timber production cost from the tree felling at PT Inhutani II.

Sampling plots for the research	Tree No.	Machine expenses (IDR/hour)	Timber productivity from the tree felling (m^3/hour)	Wages expenses (IDR/m^3)	Cost of timber production from tree felling (IDR/m^3)
I	1	35,643.3	37.855	13,000	13941.57
	2	35,643.3	40.455	13,000	13881.06
	3	35,643.3	39.106	13,000	13911.45
II	4	35,643.3	35.991	13,000	13990.34
	5	35,643.3	36.220	13,000	13984.08
	6	35,643.3	40.833	13,000	13872.90
	7	35,643.3	42.547	13,000	13837.74
	8	35,643.3	34.909	13,000	14021.03
	9	35,643.3	36.398	13,000	13979.27
	10	35,643.3	35.752	13,000	13996.96
III	11	35,643.3	40.539	13,000	13879.23
	12	35,643.3	38.789	13,000	13918.90
	13	35,643.3	39.704	13,000	13897.73
	14	35,643.3	38.247	13,000	13931.92
	15	35,643.3	40.244	13,000	13885.68
	16	35,643.3	40.804	13,000	13873.52
	17	35,643.3	43.379	13,000	13821.67
IV	18	35,643.3	37.836	13,000	13942.05
	19	35,643.3	43.108	13,000	13826.84
	20	35,643.3	40.423	13,000	13881.76
	21	35,643.3	41.022	13,000	13868.88
	22	35,643.3	38.541	13,000	13924.82
	23	35,643.3	36.705	13,000	13971.07
	24	35,643.3	39.289	13,000	13907.21
V	25	35,643.3	41.160	13,000	13865.97
	26	35,643.3	43.078	13,000	13827.41
	27	35,643.3	37.949	13,000	13939.24
	28	35,643.3	42.984	13,000	13829.22
	29	35,643.3	40.702	13,000	13875.71
	30	35,643.3	33.797	13,000	14054.63
VI	31	35,643.3	38.414	13,000	13927.87
	32	35,643.3	38.154	13,000	13934.20
	33	35,643.3	38.735	13,000	13920.18
	34	35,643.3	37.725	13,000	13944.82
	35	35,643.3	38.677	13,000	13921.56
	36	35,643.3	35.375	13,000	14007.58
	37	35,643.3	34.230	13,000	14041.29
VII	38	35,643.3	42.286	13,000	13842.91
	39	35,643.3	32.759	13,000	14088.05

	40	35,643.3	37.505	13,000	13950.36
	41	35,643.3	39.376	13,000	13905.20
	42	35,643.3	35.189	13,000	14012.91
	43	35,643.3	38.699	13,000	13921.04
VIII	44	35,643.3	32.285	13,000	14104.02
	45	35,643.3	40.859	13,000	13872.35
	46	35,643.3	40.776	13,000	13874.12
	47	35,643.3	34.367	13,000	14037.14
	48	35,643.3	37.518	13,000	13950.03
	49	35,643.3	35.739	13,000	13997.32
	50	35,643.3	35.660	13,000	13999.53
IX	51	35,643.3	34.534	13,000	14032.12
	52	35,643.3	41.039	13,000	13868.52
	53	35,643.3	36.007	13,000	13989.90
	54	35,643.3	37.866	13,000	13941.30
	55	35,643.3	36.104	13,000	13987.24
	56	35,643.3	36.690	13,000	13971.47
X	57	35,643.3	37.489	13,000	13950.77
	58	35,643.3	41.200	13,000	13865.13
	59	35,643.3	38.918	13,000	13915.86
	60	35,643.3	35.945	13,000	13991.61
Sum		2,138,598	2,298.486	780,000	836106.29
Average		35,643.3	38,308	13,000	13935.10

Table 3 shows that the average cost of timber production from the tree-felling reached IDR 13,935.10/m^3, with the machine cost worth IDR 35,643.3/hour and the average felling productivity equal to 38.3058 m^3/hour. Average timber production that resulted from this study was greater than the results by [13], who performed his research regarding felling productivity in natural forest in North Kalimantan using Stihl 070's chainsaw as well. Meanwhile, average tree-felling productivity that implemented the RIL (Reduced Impact Logging)'s techniques corresponded to 45.56 m^3/hour with the average cost of tree-felling productivity worth IDR 956.58/m^3. Such high cost of felling productivity in this research was caused by several factors, such as: 1) Low average felling productivity as achieved: tree-felling productivity affected timber-production cost, whereby the greater the felling productivity, then the lower would be the production cost; 2) The topography condition of the research area which was very steep could slow down the tree-felling process, thereby prolonging the felling duration; and 3) Despite using chainsaw with the same type, in fact, the skills afforded by the operators differed from one to others.

3.3 Timber utilization efficiency (TUE)

Optimation of wastes generated from timber harvest presents one of the several indicators toward sustainable forest management. That inducement became a reference for PT Inhutani II's attempt to minimize timber harvesting wastes by utilizing them, based on their dimension. Tree-felling wastes which were measured in this research pertained to the residues or portion of the felled trees regarded in as of no-economic values in the process of felling production and therefore were just abandoned on sites after the logging operation finished. Those wastes related to the TUE very closely. The average TUE at PT Inhutani II is presented in Table 4.

Table 4 Timber Utilization Efficiency (TUE) at PT Inhutani II.

Sampling plots for the research	Tree No.	Total volume of timber (m^3)	Volume of generated wastes (m^3)	TUE
I	1	9.984	0.520	0.95
	2	11.973	0.510	0.96
	3	9.557	0.433	0.95
II	4	8.371	1.292	0.85
	5	8.470	1.407	0.83
	6	6.345	0.901	0.86
	7	7.101	1.286	0.82
	8	4.638	0.682	0.85
	9	7.404	0.185	0.97
	10	4.439	0.268	0.94
III	11	3.269	0.566	0.83
	12	4.461	0.453	0.90
	13	5.205	0.242	0.95
	14	10.416	0.790	0.92
	15	13.983	1.843	0.87
	16	6.494	1.053	0.84
	17	4.596	0.258	0.94
IV	18	9.891	0.306	0.97
	19	7.747	0.203	0.97
	20	4.603	0.156	0.97
	21	12.794	0.829	0.94
	22	4.905	0.152	0.97
	23	2.585	0.077	0.97
	24	4.943	0.163	0.97
V	25	6.679	0.437	0.93
	26	5.119	0.452	0.91
	27	6.504	0.179	0.97
	28	16.357	1.241	0.92
	29	8.288	0.758	0.91
	30	3.793	0.301	0.92
VI	31	12.839	1.251	0.90
	32	2.979	0.181	0.94
	33	8.047	0.622	0.92
	34	4.729	0.768	0.84
	35	8.168	0.690	0.92
	36	4.519	0.687	0.85
	37	2.604	0.150	0.94
VII	38	8.150	0.327	0.96
	39	3.165	0.162	0.95
	40	6.524	0.273	0.96

	41	10.341	0.366	0.96
	42	10.699	0.670	0.94
	43	4.128	0.258	0.94
VIII	44	4.199	0.110	0.97
	45	7.197	0.319	0.96
	46	5.971	0.194	0.97
	47	3.758	0.150	0.96
	48	3.899	0.147	0.96
	49	7.714	0.566	0.93
	50	11.021	0.620	0.94
IX	51	3.735	0.282	0.92
	52	15.646	1.966	0.87
	53	9.621	0.439	0.95
	54	10.549	1.019	0.90
	55	9.583	0.256	0.97
	56	10.925	0.468	0.96
X	57	5.830	1.082	0.81
	58	4.063	0.492	0.88
	59	8.456	0.608	0.93
	60	8.300	1.111	0.87
Sum		438.269	34.176	55.33
Average		7.305	0.569	**0.92**

Table 4 reveals that average total volume of harvested timber and volume of the generated wastes were consecutively 7.305 m^3 and 0.569 m^3, thereby leading to as much 0.92 or 92% of the timber as utilized, while the rest (8%) presented as the generated wastes which were abandoned. From the 10 plots prepared as the research sample, it disclosed that the range of timber utilization was 0.81-0.97. This indicated that the lowest TUE occurred at plots II and X which reached 0.82 and 0.81, respectively. When assessed from their topography, those two plots belonged to the steep areas, thereby causing a lot of tree portions to sustain technical damages due to the high number of stumps abandoned on sites, broken stems, and some of the felled trees accidentally falling down into deep ravines.

The abandoned wastes in this research amounted to 8%, with average timber production from those wastes per year equal to 3,275 m^3 [14]. The efficiency of timber utilization from such particular amount of the generated wastes (8% implied that the Company (PT Inhutani II) would gain additional profit as production increase which corresponded to 8% multiplied by 3,275 m^3 or equal to 262 m^3/year. Assuming that the timber price was IDR 1,000,000 per m^3 and the feasible profit for the Company corresponded to 20% (or equal to IDR 200,000 / m^3), and therefore the Company itself would gain as much 262 m^3/year multiplied by IDR 200,000/m^3 or equal to IDR 52,400,000/year as an additional profit. Such high additional profit would supposedly occur just at this study that covered an area of 2,041.04 ha. Accordingly, when accounted for the entire area of production forests managed by the Company, then a significant reduction would occur in the amount of generated wastes and therefore deserve thorough consideration.

TUE related closely to the annual allocated production (AAP) in the forest enterprise (entrepreneurship). The AAP value could be calculated using the related formula: V = L multiplied by P by 0.92 or 0.80, where V = volume of the logged/harvested timber that could be felled per year: L (etat area) = actual area of the forest from which its tree stands could be felled per year; P = potency of forest logging in accordance with the diameter limit related to the forest function (m^3/ha); 0.92 =

efficiency of timber utilization as acquired in this research (Table 4); and 0.80 = security factor. Using the data which were quoted from the AWP (Annual Working Plan) book in the year 2016 owned by the Company (PT Inhutani II), then the forest area that could be felled reached 2,041.04 ha/year; and the logging potency in accordance with the diameter limit corresponded to 112.555 m^3 per ha. In this way, therefore the AAP value would be 2,041.04 multiplied by 112.555 by 0.92 or 0.80 = 169,080.73 m^3/year. Further, if the efficiency value for timber utilization or exploitation factor complied with the regulation enacted by the Ministry of Forestry (MOF) which corresponded to 0.70, then the AAP value would be 128,648.39 m^3/year. As such, a difference would occur in AAP values between efficiency value of timber utilization as stipulated by the MOF and the efficiency value obtained in these research results, which would be 169,080.73 - 128,648.38 or equal to 40,432.35 m^3/ year.

Research results by [15] regarding the exploitation factor in East Kalimantan and Central Kalimantan which corresponded to 0.82 would bring about an impact on the national AAP value from the average value as much as 8.756 million m^3/year to the other average value (10.052 m^3/year) or being equivalent to additional production of round timber approximately 1.077 million m^3 per year.

Results of this research were greater than those conducted by [2] a natural forest area in West Sumatera and Central Kalimantan, which disclosed that TUE in those two respective areas reached consecutively 0.75 and 0.74. This implied from their research that the degree of timber utilization starting from the standing trees until they were removed from the forest corresponded to 74-75%, while the residual (24-25%) signified as wood wastes abandoned in the forest.

3.4 Residual stands damage

Activities of the tree felling could bring about various kinds of damages to the residual stands. According to Elias (2008), kinds of the damaged that occurred to the abandoned residual stands could be among others damaged canopies, shattered barks, ruptured stems, broken stems, leaning stands, trees felled down, and damaged buttresses. Such damages were due to being hit by the felled trees, either hit by canopies, stems or hit by the stem of the felled down trees. In relevant, the average damages that occurred to the residual tree stands at PT Inhutani II, as conducted in this research are disclosed in Table 5.

Table 5 Average residual stands damage at PT Inhutani II.

Plot No.	Trees / ha*)	felled trees	Residual stands damage					
			Canopy	Broken	Stem injury	Felling down	Sum	
		(trees)	(trees)	(trees)	(trees)	(trees)	(trees)	(%)
Plot 1	49	3	1	2	0	1	4	8.70
Plot 2	57	8	2	2	1	3	8	16.33
Plot 3	55	7	2	2	2	1	7	14.58
Plot 4	55	7	3	2	1	1	7	14.58
Plot 5	53	6	1	2	0	3	6	12.77
Plot 6	57	7	0	3	1	3	7	14.00
Plot 7	49	5	2	2	1	1	6	13.64
Plot 8	57	8	2	3	1	2	8	16.33
Plot 9	50	6	2	1	2	1	6	13.64
Plot 10	49	3	1	1	1	2	5	10.87
Average	53.1	6.0	1.6	2	1	1.8	6.4	**13.54**

Note : *) Number of tree stands with their diameter (Ø) equal to or greater than 20 cm before felling

Table 5 reveals that the average number of tree stands with diameter greater than 20 cm corresponded to 53 trees per ha, with the number of the felled trees in the range of 3-8 trees/ha or on average of 6 trees/ha. The tree felling inflicted the damages on the abandoned residual stands, which reached 6.4 trees/ha on average (13.54%). This implied for every one tree being felled, it left with roughly one residual tree stands damaged. In general, the damages to the residual stands, which could comprise damaged canopy, ruptured tree stems, stem injury, and felling down, revealed specifically as many as 1.6 trees with destroyed canopies (1.6/6.4 x 100 % = 25%), 2 trees with ruptured stem (2/6.4 x 100% = 31.25%), 1 tree with injured stem (1/6.4 x 100% = 15.63%), and 1.8 trees felling down (1.8/6.4 x 100% = 28.13%).

Experiment results in this research brought about the residual stands damage (13.54%) which was lower than those found by [3] the latter revealed that the residual stands damage at the trees due to felling in the tropical peat natural forest reached 20%. Meanwhile, research results by [16] disclosed that the tree felling in a natural forest in Iran numbered to 5.1 trees per ha, in that as much 17.3 m^3/ha. As such, the damages that occurred to residual stands corresponded to 9.8 trees.

4. Conclusion and suggestions

The tree-felling techniques as implemented typified as the felling techniques which brought about low impacts or so-called RIL (Reduced Impact Logging) technique. Average tree-felling productivity reached 38.308 m3/hour, with average felling production worth Rp 13,897.44/m3. Average TUE corresponded to 92%, with the concurrently generated wastes as abandoned equal to 8%. The average of residual stands damage due to the tree felling reached 13.54%, which comprised of canopy damage, ruptured/broken tree stem, stem injury, and tree felling down. The tree-felling techniques adopting the RIL (Reduced Impact Logging) techniques indicatively were able to come up with convenient results/expectation (e.g. high felling productivity, low felling cost, high TUE, and low portion of the generated wastes) provided by paying meticulous attention to the field condition (e.g. steepness of topography, types of forests, dimension of the logged, buttresses/canopy/branch-free stems, and residual stands damage) and the affecting factors (e.g. chainsaw machine, type and specification, operators, their health, and skills); and therefore this deserves thorough consideration in the commercial implementation.

Acknowledgement

The authors wish to express sincere gratitude and profound appreciation to Dr. Ratih Damayanti, who has thoroughly and meticulously reviewed this manuscript (scientific writing) as capably as she could.

References

[1] Sitohang W R M, Muhdi and Afifuddin Y 2016 https://jurnal.usu.ac.id/index.php/PFJS/article/view/14163. Access on April 30, 2019.
[2] Matangaran J R, Partiani T and Purnamasari D R 2013 *J. Bumi Lestari* **13** 384-393
[3] Suwarna U, Matangaran J R and Harmawan F 2014 *J. Manusia dan Lingkungan* **21** 83-9
[4] Directorate General of Forest Enterpreneurship 1994 *Pedoman dan Petunjuk Teknis Pelaksanaan Sistem Silvikultur Tebang Indonesia (TPTI)* (Jakarta: Departemen Kehutanan)
[5] Elias 1998 *Reduced Impact Timber Harvesting in the Indonesian Selective Cutting and Planting System* (Bogor: IPB Press)
[6] Idris M M and Sukanda 2012 *J. Penelitian Hasil Hutan* **30** 269-278

[7] Ruslandi 2013 *Petunjuk Teknik Penerapan Pembalakan Berdampak Rendah Karbon (RIL-C) pada Izin Usaha Pemanfaatan Hasil Hutan Kayu Hutan Alam (IUHHK-HA)* (Jakarta: The Nature Conservancy)
[8] FAO 1992 *Cost control in forest harvesting and road construction (FAO Forestry Paper No 99)* (Rome: FAO)
[9] Thaib J and Suhartana S 1991 *J. Penelitian Hasil Hutan* **9** 144-9
[10] Behjou F K, Majnounian B, Dvorak J, Namiranianl M, Saeed A and Feghhi J 2009 *J. of For.Sc.* **55** 96-100
[11] Akay E, Erdas O, Buyuksakall H and Akar D S 2015 *Proc.of the 48th FORMEC Symp.* Ed C Kanzian, G Erber and M Kuhmaier (Austria: Institute of Forest Engineering) 77-80
[12] Suhartana S and Yuniawati 2006 *Peronema For Sci J* **2** 37-44
[13] Muhdi 2016 *Int J of Sci and Res* **5** 141-3
[14] Inhutani II 2015 *Rencana Kerja Tahunan IUPHHK-HA Inhutani II Unit Sei Tubu, Kabupaten Malinau, Provinsi Kalimantan Utara* (Malinau: Inhutani II)
[15] Idris M M, Dulsalam, Sukanda and Soenarno 2012 *Proc Ekspose Hasil-Hasil Penelitian* (Bogor: Pusat Litbang Hasil Hutan)
[16] Behjou F K and Mollabashi O G 2012 *BioResources* **7** 4867-74

IOP Conf. Series: Earth and Environmental Science **359** (2019) 012008 doi:10.1088/1755-1315/359/1/012008

Characteristics of TEMPO-Oxidized Cellulose Nanofibers from Oil Palm Empty Fruit Bunches Produced by Different Amounts of Oxidant

N. Hastuti[1,2*], K. Kanomata[1] and T. Kitaoka[1]

[1] Graduate School of Bioresource and Bioenvironmental Sciences, Kyushu University, Motooka 744, Nishi-ku, Fukuoka, Japan, 819-0395

[2] Forest Products Research and Development Center, Ministry of Environment and Forestry, Republic of Indonesia, Jalan Gunung Batu 5, Bogor, West Java, Indonesia, 16610

*email: novienov3@yahoo.com

Abstract. The use of oil palm waste as a source of raw materials for nanocellulose production is a promising approach for turning agricultural waste into valuable products. This study aimed to characterize cellulose nanofibers (CNFs) derived from oil palm empty fruit bunches (OPEFB) produced by 2,2,6,6-tetramethylpiperidine 1-oxyl (TEMPO)/NaBr/NaClO. TEMPO-mediated oxidation was carried out at pH 10 by applying different amounts of oxidant: 10, 15 and 20 mmol/g-cellulose at room temperature for 2 h, where the resultant CNFs were denoted as T-10, T-15 and T-20, respectively. Characterization results revealed that the carbon content of each CNF decreased as compared to the starting pulp, from 40% to *ca.* 37%. Microscopic analysis showed that the aspect ratios of obtained CNFs remained almost unchanged by increasing the amounts of oxidant, which were ca. 40. The crystallinity indices (CrIs) ranged from 34 to 55%. The carboxy contents of CNFs increased by adding higher amounts of oxidant, but the thermal resistance properties were slightly decreased. High-performance nanomaterials such as TEMPO-oxidized CNFs were successfully prepared to show good characteristics, from low-quality biomass waste such as OPEFB.

1. Introduction

Indonesia is acknowledged as the largest producer of palm oil in the world. Oil palm plantation expanded 100-fold in the last half-century, most of them locating on Sumatra, Kalimantan and Sulawesi island [1,2]. After isolation of crude palm oil and palm kernel oil, a large amount of residual oil palm empty fruit bunches (OPEFB) are left as the main waste in the plantation site without further processing [3]. Recently, OPEFBs have attracted much attention as cellulose resource because they contain about 50-60% of cellulose by weight [4,5]. Utilization of OPEFB, hugely abundant agricultural waste in Indonesia, as cellulose source hopefully addresses the environmental issue by turning it into valuable products.

Among cellulose-derived materials, nanocellulose has attracted great interest in recent years because of its promising properties such as mechanical strength and thermal stability. Applications of nanocellulose are now in intensive study in academia and industry, including nanocomposites, gas-barrier films, and electronic devices [6]. Nanocellulose has been determined as cellulose nanocrystals (CNCs) and cellulose nanofibrils (CNFs) based on their morphological characteristics. Furthermore, CNCs are produced by chemical treatments, followed by mechanical disintegration, and CNFs are obtained by either mechanical or chemical treatments. CNCs are highly crystalline with a length less than 500 nm and CNFs are consist of mixture amorphous and crystalline with length of up to several micrometers [7]. One of the obvious methods to produce CNFs is done by 2,2,6,6-tetramethylpiperidine 1-oxyl (TEMPO)-mediated oxidation. TEMPO has been widely studied for catalytic and selective oxidation of primary hydroxy groups under aqueous condition. In case of CNFs production by applying TEMPO/NaBr/NaClO system at pH 10, C6 primary hydroxy groups of cellulose effectively converted to sodium carboxylate groups, while maintaining fibrous morphologies such as their crystallinity [8].

In this study, we produce cellulose nanofibers from oil palm empty fruit bunches (OPEFB) by 2,2,6,6-tetramethylpiperidine 1-oxyl (TEMPO)-mediated oxidation. This process was effective to produce cellulose nanofibers with the surface-modified nanocellulose with various functional groups, long and individualized cellulose fibrils dispersed in water [8,9]. This method also effectively catalyzed the regioselective conversion on primary hydroxy groups of C6 of pulp fibers [10]. Different amounts of oxidant, sodium hypochlorite (NaClO), were applied in the oxidation process. The obtained characteristics of TOCNs from OPEFB are potential to be applied in producing nanocellulose-based biomaterials.

2. Materials and method

2.1. Materials

Bleached kraft pulp of oil palm empty fruit bunches (OPEFB) was kindly supplied from Biomaterial Research Institute, Indonesian Institute of Sciences (Bogor, Indonesia). 2,2,6,6-Tetramethylpiperidine 1-oxyl (TEMPO), sodium bromide, sodium hypochlorite, and sodium borohydride were purchased from Sigma-Aldrich (Tokyo, Japan) and used without further purification. The water used in this study was purified with an Arium Ultrapure Water System (Sartorius Co., Ltd., Tokyo, Japan).

2.2. Preparation of TOCNs

The bleached kraft pulp of OPEFB was soaked in a 0.01 M HCl solution for 30 min for demineralization. TOCN was prepared by TEMPO/NaBr/NaClO system at pH 10 according to the literature [8,9]. In brief, a 2.5-g of dry weight of demineralized OPEFB pulp (about 85% of cellulose content) was suspended in water (250 mL) containing TEMPO (16 mg/g-cellulose) and NaBr (100 mg/g-cellulose). The oxidation reaction was initiated by adding 2 M NaClO aq as oxidant (10, 15, 20 mmol/g-cellulose). The pH of suspension was maintained at 10 by adding 0.5 M aqueous NaOH with a pH titrator (Mitsubishi Chemical Analytech, Yamato, Japan) during the reaction. After 2 h, the reaction was quenched by adding ethanol (2 mL), followed by addition of $NaBH_4$ (100 mg/g-cellulose), and the resultant mixture was further stirred for 1 h. The obtained suspension was thoroughly washed using deionized water by centrifugation and then sonicated by an ultrasonic homogenizer for 5 min (US-300E, Nihonseiki Ltd., Tokyo, Japan). Residual unfibrillated fibers were removed by further centrifugation at 12000 *g* for 10 min. Obtained TOCNs, denoted as T-10, T-15, and T-20 corresponding to the amount of NaClO used in the oxidation, were kept at 4°C until further use.

2.3. Characterization of TOCNs

Elemental analysis was performed with an Organic Micro Analyzer CHN CORDER MT-6 (Yanaco Ltd., Tokyo, Japan). The surface morphology of original OPEFB pulp was observed using a scanning electron microscope (SEM SU-3500, Hitachi Ltd., Tokyo, Japan) at the Center of Advanced Instrumental Analysis, Kyushu University. Samples were mounted on carbon tape without any coating, and the machine was operated at an acceleration voltage of 15 kV and pressure of 30 Pa. The length and width of TOCNs were measured by a transmission electron microscopy (TEM) (JEM 2100-HC, JEOL Ltd., Tokyo, Japan) operated at an acceleration voltage of 120 kV at the Ultramicroscopy Research Center, Kyushu University. The size was measured using Image-J version 1.51s. Structural analysis was observed by X-ray diffraction (XRD), Fourier Transform Infrared (FTIR) and thermogravimetric analysis (TGA) of TOCNs were conducted as described in our previous work [11]. The carboxylate contents were measured by conductometric titration [8].

3. Results and discussion

3.1. Elemental Analysis and Carboxylate Content

Elemental analysis revealed that the carbon content (C%) of obtained TOCNs decreased as compared to the starting pulp (**Table 1**). Higher amounts of oxidant applied in this study revealed that the C% almost unchanged by increasing the oxidant, indicating that TEMPO-mediated oxidation selectively occurred on the primary hydroxy groups of cellulose and did not affect cellulose polymorph. The increase of carboxylate content by increasing the amount of primary oxidant (NaClO) indicated that the carboxylate and aldehyde groups were formed more in the water-insoluble fractions by the oxidation. However, at certain amount of NaClO addition, the carboxylate content will not exceed a certain level even by the excess of NaClO addition [8].

Table 1. Carbon and carboxylate contents

Sample	Carbon content (C%)	Carboxylate content (mmol/g)
Raw pulp	40.13	n.d.
T-10	38.32	1.08
T-15	38.67	1.34
T-20	37.77	1.50

Remarks: n.d. = not determined ; T-10 produced by 10 mmol/g-cellulose of NaClO; T-15 produced by 15 mmol/g-cellulose of NaClO; T-20 produced by 20 mmol /g-cellulose of NaClO

3.2. Morphology and size of TOCNs

Morphology of raw pulp and obtained TOCNs was shown in **Figure 1**. Microscopy analysis of original pulp showed that TEMPO-mediated oxidation, followed by ultrasonication successfully afforded nanofibrillated fibers from OPEFB pulp. The length of resultant TOCNs was 100–291 nm, and the width was 3–9 nm. The aspect ratios were 40 ± 22 for T-10, 41 ± 19 for T-15, and 41 ± 14 for T-20 (**Fig. 1**). Notably, the aspect ratios of resultant TOCNs were almost unchanged by increasing the amount of oxidant. These results were in good accordance that the carboxylate groups are not formed inside the cellulose crystallite during TEMPO-mediated oxidation [8]. The oxidation occurred only on C6 primary hydroxyl groups on the surfaces of fibrils in cellulose fibers, therefore no significant changes in the fibrous morphology [9]. The aspect ratios of obtained TOCNs were much higher than those reported in the previous study, in which TOCNs were produced from microcrystalline cellulose (MCC) of OPEFB [12]. The results showed that TEMPO-oxidized cellulose nanofibrils (TOCNs) were successfully obtained from low quality biomass waste with a relatively high content of carboxy groups and aspect ratios

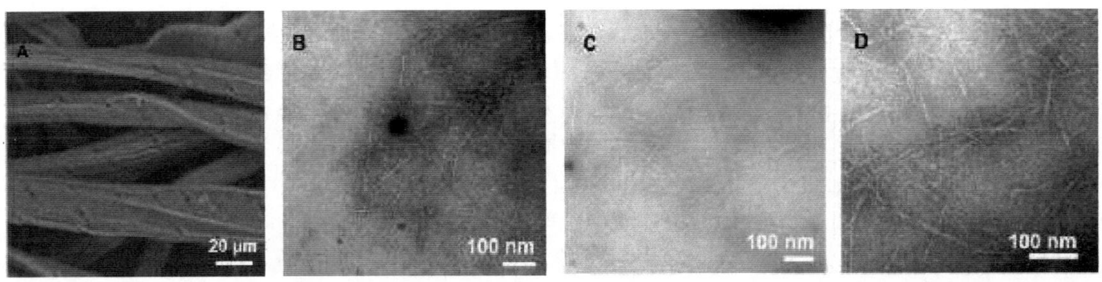

Figure 1. SEM image of raw OPEFB pulp (A) and TEM images of the resultant TOCN T-10 (B), T-15 (C), and T-20 (D)

3.3. Structural Analysis

The FTIR and XRD analyses of the starting pulp and resultant TOCNs are shown in **Figure 2**. In FTIR spectra, the peaks at 1618 cm^{-1} of resultant TOCNs are attributed to carboxylate groups, and small absorption bands at 1720 cm^{-1} are C=O groups in the protonated carboxyl groups [13]. This absorption band was not found on the starting pulp spectra. The FTIR spectra justified that during the oxidation process, a significant amount of carboxylate groups were introduced into nanocellulose surface [9].

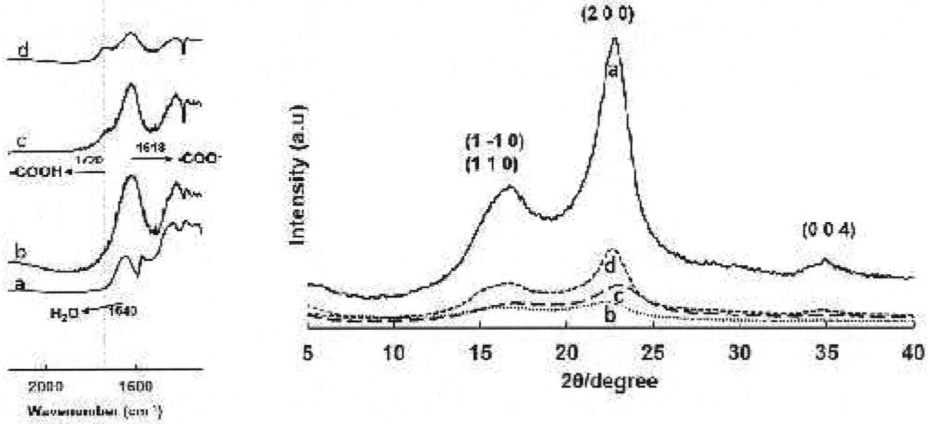

Figure 2. FTIR spectra and XRD patterns of: (a) raw pulp of OPEFB (b) T-10; (c) T-15 and (d) T-20

XRD profiles showed that crystallinity indices (Cr.I) of resultant TOCNs decreased after the oxidation process (**Fig. 2**). It might be caused by the regio-selective conversion of primary hydroxy groups at the C6 position of raw pulp (cellulose) to carboxylate ones by having sodium glucuronosyl units and made the structure of raw cellulose turn to disordered structure [9,14]. The 200 and 110 diffraction peak positions of TOCNs slightly changed after oxidation, it may indicate the introduction of carboxylate and aldehyde groups on the surface of cellulose I [8]. The Cr.I of raw pulp was 61% and for T-10, T-15 and T-20 were: 34%, 46%, and 55%, respectively. The higher amount of oxidant added in the oxidation slightly changed the Cr.I. It was assumed that the crystalline region slightly changed with the addition of oxidant. Presumably, the carboxylate groups were mostly present on the crystal surface and in disordered region [8,10]. The Cr.I values of resultant TOCNs were lower than the previous study conducted by Rohaizu and co-author (60.5%). This lowering is affected by the difference of starting cellulose source, whereas microcrystalline cellulose of OPEFB was used in the previous study which was relatively highly crystalline [12].

Thermal analysis was performed by thermogravimetry (TG) and differential thermogravimetry (DTG); the onset temperature (T_{on}) and the maximum degradation temperature (T_{max}) as shown in **Figure 3**. Thermal stability of resultant TOCNs was sharply decreased compared to the starting pulp, because carboxylate groups trigger rapid decomposition of polysaccharides at a lower temperature [16]. The amount of oxidant in the preparation of TOCN did not affect the thermal properties of resultant TOCNs. On the other hand, the char residue remaining after 500°C of resultant TOCNs was much higher compared to that with the starting pulp. It indicates that carboxylate groups on TOCNs have a flame retardant effect [12]. These thermal properties parameters were close to TOCNs made of MCC OPEFB but lower than TOCNs made from corn husk pulp [12,15].

IOP Conf. Series: Earth and Environmental Science **359** (2019) 012008 doi:10.1088/1755-1315/359/1/012008

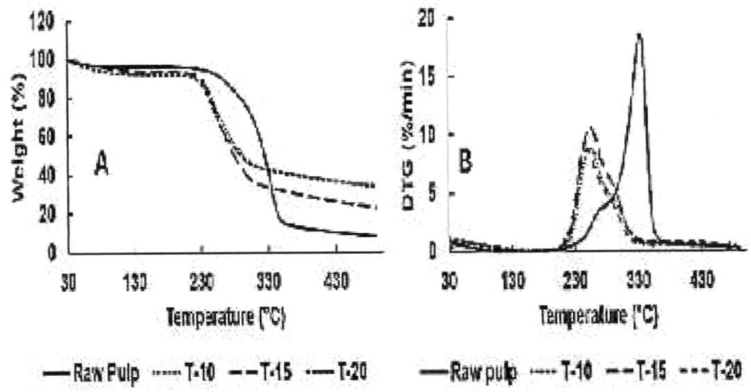

C

Sample	T_{on} (°C)	T_{max} (°C)
Raw pulp	252.2	359.2
T-10	220.1	251.2
T-15	215.4	252.1
T-20	211.5	250.1

Figure 3. TG curves (A); DTG curves (B); and thermal properties of raw pulp and resultant TOCNs (C)

4. Conclusion

TEMPO-oxidized cellulose nanofibers (TOCNs) were successfully produced from low quality of agricultural residue, oil palm empty fruit bunches (OPEFB). The differences in applying the amount of primary oxidant, NaClO, were not affected by the TOCNs morphologies which were observed by microscopy and XRD analysis. Higher amount of oxidant slightly increased the carboxylate content and induced lower thermal stability of TOCNs OPEFB than raw pulp. High aspect ratios and the presence of carboxylate groups of TOCNs are promising properties for further application such as reinforcing agents in plastics and nanoparticle anchor in catalytic reactions.

Acknowledgements

This research was supported by an Advanced Low Carbon Technology Research and Development Program from the Japan Science and Technology Agency (T.K.) and by a Research Fellowship for Young Scientists from Japan Society for the Promotion of Science (K.K.). N.H. convey a great thanks to the Ministry of Education and Culture, Republic of Indonesia, for providing Unggulan Scholarship. The authors also thank Prof. Dr. Syuntaro Hiradate and Prof. Dr. Yuki Mori for performing XRD analysis.

References

[1] Harsono, L.T. Mulyantara, A.T. Rizaluddin, A. Nakagawa-izumi, H. Ohi, K. Nakamata, Properties of fibers prepared from oil palm empty fruit bunch for use as corrugatting medium and fiberboard, Japan TAPPI J. 69 (2015) 67–77.

[2] R. Moreno-Peñaranda, A. Gasparatos, P. Stromberg, A. Suwa, A.H. Pandyaswargo, J.A. Puppim de Oliveira, Sustainable production and consumption of palm oil in Indonesia: What can stakeholder perceptions offer to the debate?, Sustain. Prod. Consum. 4 (2015) 16–35. doi:10.1016/j.spc.2015.10.002.

[3] H. Harsono, A.S. Putra, R. Maryana, A.T. Rizaluddin, Y.Y. H'ng, A. Nakagawa-izumi, H. Ohi, Preparation of dissolving pulp from oil palm empty fruit bunch by prehydrolysis soda-anthraquinone cooking method, J. Wood Sci. 62 (2016) 65–73. doi:10.1007/s10086-015-1526-3.

[4] M.S. Siti Aisyah, Y. Uemura, S. Yusup, The Effect of Alkaline Addition in Hydrothermal Pretreatment of Empty Fruit Bunches on Enzymatic Hydrolysis Efficiencies, Procedia Chem. 9 (2014) 151–157. doi:10.1016/j.proche.2014.05.018.

[5] Z.A.Z. Azrina, M.D.H. Beg, M.Y. Rosli, R. Ramli, N. Junadi, A.K.M.M. Alam, Spherical nanocrystalline cellulose (NCC) from oil palm empty fruit bunch pulp via ultrasound assisted hydrolysis, Carbohydr. Polym. 162 (2017) 115–120. doi:10.1016/j.carbpol.2017.01.035.

[6] S. Mondal, Preparation, properties and applications of nanocellulosic materials, Carbohydr. Polym. 163 (2017) 301–316. doi:10.1016/j.carbpol.2016.12.050.

[7] N. Grishkewich, N. Mohammed, J. Tang, K.C. Tam, Recent advances in the application of cellulose nanocrystals, Curr. Opin. Colloid Interface Sci. 29 (2017) 32–45. doi:10.1016/j.cocis.2017.01.005.

[8] T. Saito, A. Isogai, TEMPO-mediated oxidation of native cellulose. The effect of oxidation conditions on chemical and crystal structures of the water-insoluble fractions, Biomacromolecules. 5 (2004) 1983–1989. doi:10.1021/bm0497769.

[9] T. Saito, S. Kimura, Y. Nishiyama, A. Isogai, Cellulose nanofibers prepared by TEMPO-mediated oxidation of native cellulose, Biomacromolecules. 8 (2007) 2485–2491. doi:10.1021/bm0703970.

[10] A. Isogai, T. Saito, H. Fukuzumi, TEMPO-oxidized cellulose nanofibers, Nanoscale. 3 (2011) 71–85. doi:10.1039/c0nr00583e.

[11] N. Hastuti, K. Kanomata, T. Kitaoka, Hydrochloric Acid Hydrolysis of Pulps from Oil Palm Empty Fruit Bunches to Produce Cellulose Nanocrystals, J. Polym. Environ. 0 (2018) 0. doi:10.1007/s10924-018-1248-x.

[12] R. Rohaizu, W.D. Wanrosli, Sono-assisted TEMPO oxidation of oil palm lignocellulosic biomass for isolation of nanocrystalline cellulose, Ultrason. Sonochem. 34 (2017) 631–639. doi:10.1016/j.ultsonch.2016.06.040.

[13] M. Shimizu, H. Fukuzumi, T. Saito, A. Isogai, Preparation and characterization of TEMPO-oxidized cellulose nanofibrils with ammonium carboxylate groups, Int. J. Biol. Macromol. 59 (2013) 99–104. doi:10.1016/j.ijbiomac.2013.04.021.

[14] B. Puangsin, Q. Yang, T. Saito, A. Isogai, Comparative characterization of TEMPO-oxidized cellulose nanofibril films prepared from non-wood resources, Int. J. Biol. Macromol. 59 (2013) 208–213. doi:10.1016/j.ijbiomac.2013.04.016.

[15] X. Yang, F. Han, C. Xu, S. Jiang, L. Huang, L. Liu, Z. Xia, Effects of preparation methods on the morphology and properties of nanocellulose (NC) extracted from corn husk, Ind. Crops Prod. 109 (2017) 241–247. doi:10.1016/j.indcrop.2017.08.032.

IOP Conf. Series: Earth and Environmental Science **359** (2019) 012009 doi:10.1088/1755-1315/359/1/012009

The utilization of activated carbon as micronutrients carrier in slow release fertilizer formulation

D. A. Prastiwi[1*], B Sumawinata[2], Iskandar[2] and G Pari[3]

[1]Cilegon Institute of Chemical Analyst, Banten, Indonesia.
Jl. KH. Wasyid No. 6 Jombang Wetan, Cilegon – Banten, Indonesia.
[2]Soil Science and Land Resource Department, Bogor Institute of Agriculture.
[3]Forest Product Research and Development Center, Bogor, Indonesia.

*email: dina.alva@yahoo.com

Abstract. In many plantations, fertilizers were often added in high dosage and applied several times to get maximum productivity. But processes in soil caused not all nutrients in fertilizer could be uptaken by plants. Besides, the high dosage can be toxic to plants. The use of slow release fertilizer can overcome this case. This research was aimed to obtain the data of adsorptive capacity of activated charcoal to the nutrients mixed and gathering the information about nutrients release rate of fertilizer produced. The study was conducted by converting materials into activated charcoal. SRF was formulated by soaking fine activated charcoal $CuSO_4$, $FeSO_4$ or $ZnSO_4$ solution for 24 hours. Analysis conducted were the observation of surface topography, qualitative analysis using EDX, and determination of total Cu, Fe and Zn in SRF. The optimal treatment was chosen to determine its release rate by extracting fertilizer with distilled water and 2% citric acid. The results showed the activated charcoal could be used to adsorb nutrients mixed as shown by SEM and EDX results. Nutrients in the charcoal could be released slowly and could not be easily leached out.

1. Introduction

Fertilization is an effort to meet the plant's nutrient need in order to increase productivity. Sometimes, fertilizers were added in high dosage and several times to reach optimum productivity. However, processes in soil caused not all nutrients could be used optimally by plants. Fertilization in high dosage could also be toxic to the seed in its growth period.

Fertilization that carrier out several times during the planting period increase the production cost to pay the labor. This can be overcome with the use of slow release fertilizer. Slow release fertilizer is a fertilizer that contains nutrients in it as a chemical compound or has certain physical properties so that the availability is delayed shortly after the application until finally used by plants [1]. This delay can be done one of them by entering the required nutrients into a porous material that can act as a "home" for these nutrients.

This study tries to utilize activated charcoal as a nutrient carrier in producing slow release fertilizer. Alfianto's research[2] shows that activated charcoal has the potential to be used as a micro fertilizer carrier. Morphologically, activated charcoal has a pore that is very effective in binding and storing nutrients in the surrounding soil which is then released slowly according to the rate consumed by plants.

According to Sudrajat and Soleh [3], activated charcoal raw material determines the active charcoal character including its ability to adsorb a material. The raw material used in this study is bamboo dominated by macropores and coconut shells which are dominated by micropores. Based on Nurhayati [4] research, bamboo is one of the raw materials that can be used to produce activated charcoal because it has a high yield of charcoal and adsorption capacity.

Several studies have shown that hard raw materials such as coconut shell, oil palm, and coal produce active charcoal with high density which is suitable for gas absorption[5]. Raw materials that have large amounts of pores with small diameters and hard textures have a broad active surface and can be made of activated charcoal with high adsorption capacity. One of the raw materials with such properties is the coconut shell.

Activated charcoal has the ability to adsorb liquids and gases of various sizes. The elements added in this study were Cu, Fe, and Zn which are micronutrients. Micronutrients are nutrients needed by plants in small amounts. The addition of these elements in excess will be toxic to the plant. By "inserting" micro nutrients into activated charcoal, it is expected that the plant's needs for these elements can be fulfilled without worrying that the added dose will be excessive because nutrients will be released slowly according to the plants taken.

This study was aimed to identify the characteristics of activated charcoal from bamboo and coconut shell activated at temperatures of 600°C and 700°C and steamed for 90 minutes, obtain data on activated charcoal adsorption capacity on mixed nutrients, and obtain information on the nutrient release rate of the mixing activated charcoal with nutrients produced.

2. Research method

2.1. Conversion of raw material into charcoal

Charcoal production is done by using a drum furnace. Combustion is carried out at a temperature that increased gradually until it reaches a temperature of ± 500°C. After all the raw materials in the furnace are completely burned, the combustion process is stopped by closing all the paths that air passes into the furnace. The process of cooling charcoal in the furnace is carried out for ± 24 hours.

2.2. Charcoal activation and analysis of activated charcoal properties

The process of producing activated charcoal is done by storing charcoal into an electric retort and activating it with water vapor at a temperature of 600°C and 700°C for 90 minutes. The activated charcoal produced was then ground to pass the 100 mesh sieve and analyzed based on SNI 06-3730-1995[6] which included rendemen active charcoal, water content, volatile content, ash content, fixed carbon content, activated charcoal adsorption capacity against iodine solution, and adsorption capacity of activated charcoal against methylene blue solution.

2.3. Formulation of Slow Release Fertilizer

Fertilizer formulation is done by soaking fine activated charcoal powder (100 mesh) in "fertilizer solution" for ± 24 hours. The concentration of the solution used is 1N or 2N and the ratio of activated charcoal to the solution is 1: 3, 1: 5, or 1: 7. Then the immersion results are cleaned and dried. After obtaining dry powder fertilizer, the analysis was then carried out which included observation of surface topography using SEM and EDX, and analysis of total Cu, Fe and Zn content by the wet digestion method.

2.4. Testing of nutrient release rates in fertilizer

This experiment was aimed to determine the release of Cu^{2+}, Fe^{2+} and Zn^{2+} contained in fertilizers. The experiment was carried out by extracting fertilizer with distilled water and 2% citric acid for 0; 15; 30; 45 and 60 minutes, and by washing the fertilizer for 25 times, then a qualitative analysis using EDX was carried out.

3. Results and discussion

3.1. Results of characterization of active charcoal analysis

Results of characterization of charcoal analysis were presented in Table 1.

Table 1. Results of active charcoal characterization analysis

Sample	Rendemen (%)	Water content (%)	Volatile content (%)	Ash Content (%)	Fixed Carbon (%)	Iodin Adsorption Capacity (mg/g)	Methylene blue adsorption capacity (mg/g)
AB	-	6.16	16.56	6.53	71.41	256	3.89
B1	79.04	1.59	10.90	6.86	80.65	446	18.70
B2	74.64	2.14	8.95	5.67	83.24	734	21.99
ATK	-	7.27	23.69	7.34	61.69	466	23.26
T1	85.22	2.12	14.26	2.04	81.58	648	24.94
T2	81.33	1.62	13.78	3.22	81.37	760	49.63
SNI 06-3730-1995	Not required	Max. 5	Max. 25	Max. 10	Min. 65	Min. 750	Min. 120

Note: AB = Bamboo charcoal; B1 = Activated bamboo charcoal in 600°C with 90 minutes steamed; B2 = Activated bamboo charcoal in 700°C with 90 minutes steamed; ATK = Coconut shell charcoal; T1 = Activated coconut shell charcoal produced with 90 minutes steamed; T2 = Activated coconut shell charcoal produced with 90 minutes steamed

Based on the data in Table 1, it is known that the activated charcoal produced meets SNI 06-3730-1995 standards except for adsorption power on iodine and methylene blue. The low adsorption capacity indicates that the activation treatment of the material is not enough to open the pores of the material.

3.2. Results of slow release fertilizer analysis

The optimum condition of activated charcoal in absorbing nutrients is unknown, therefore an experiment was conducted by combining the concentration of fertilizer solution and comparison of charcoal with the fertilizer solution. Furthermore, the analysis was carried out using SEM to determine the topography of the charcoal surface, activated charcoal and activated charcoal that had been given immersion treatment. The results of the observations can be seen in Figure 1 and Figure 2.

Figure 1(a) and Figure 2(a) show that the pore has formed in charcoal, but part of the surface is still covered with hydrocarbon compounds and ash. The activation process causes shrinkage in the charcoal because more volatile material is released. This is also seen in the volatile content in activated charcoal are lower than charcoal (Table 1). Activation causes the formation of new micropores and damage to the micropore wall so that the diameter becomes larger. Figures 2 (b) and 2 (c, d, e) show that the pores which were initially empty on the activated charcoal become filled after being immersed. The same results are found in Figures 3 (b) and 3 (c, d, e), only the pore size in coconut shell activated charcoal is smaller than that of bamboo activated charcoal.

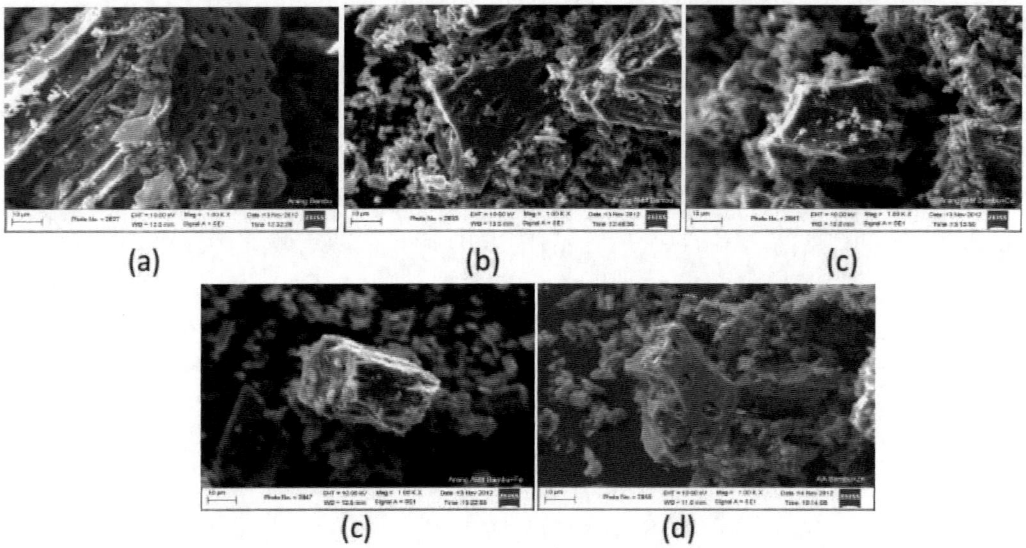

Figure 1. Surface topography of bamboo charcoal (a), bamboo activated charcoal (b), bamboo activated charcoal + Cu (c), bamboo activated charcoal + Fe (d), and bamboo activated charcoal + Zn (e) with 1000x magnification

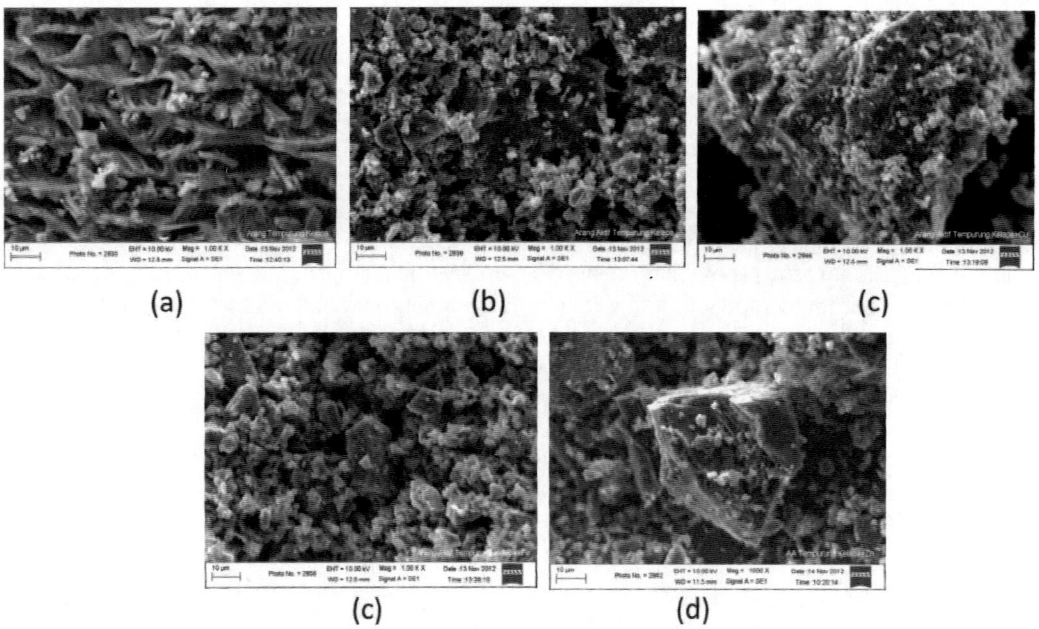

Figure 2. Surface topography of coconut shell charcoal (a), coconut shell activated charcoal (b), coconut shell activated charcoal + Cu (c), coconut shell activated charcoal + Fe (d), and coconut shell activated charcoal + Zn (e) with 1000x magnification

Although it is known that the activated charcoal pore has been filled after being given immersion treatment, further testing is needed to ensure the element that fills the charcoal. Tests are carried out qualitatively using EDX. The test results can be seen in Figure 3 and Figure 4.

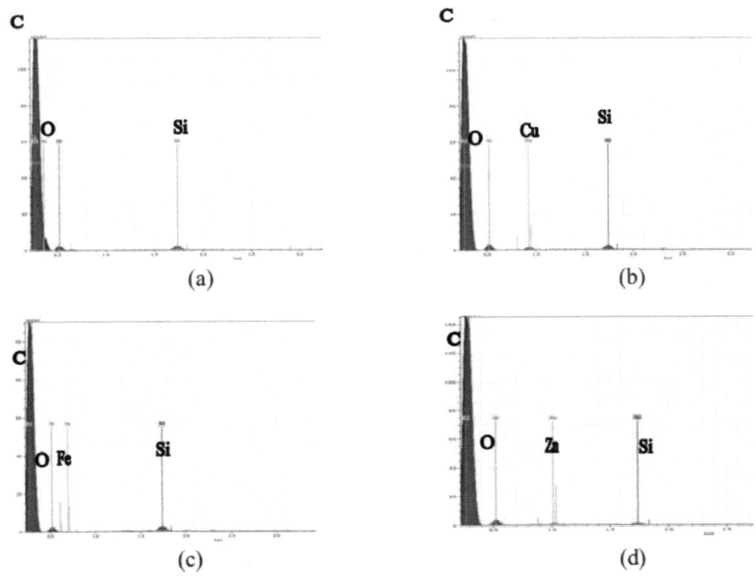

Figure 3. Observation results of EDX on bamboo activated charcoal (a), bamboo activated charcoal+Cu (b) bamboo activated charcoal+Fe (c), and bamboo activated charcoal+Zn (d)

Based on Figure 3, it is known that Cu, Fe and Zn are found in bamboo activated charcoal which has been soaked with a $CuSO_4$, $FeSO_4$, and $ZnSO_4$ solution which is then washed and dried. Qualitative analysis was also carried out on coconut shell activated charcoal and coconut shell activated charcoal given immersion treatment. Figure 4 showed that Cu, Fe, and Zn are found in coconut shell activated charcoal which has been soaked with a $CuSO_4$, $FeSO_4$, and $ZnSO_4$.

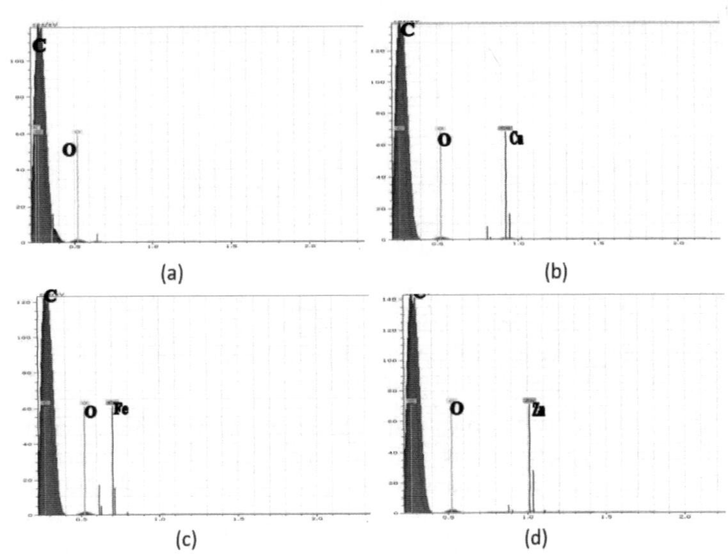

Figure 4. Observation results of EDX on Coconut Shell Activated Charcoal (a), Coconut Shell Activated Charcoal + Cu (b) Coconut Shell Activated Charcoal +Fe (c), and Coconut Shell Activated Charcoal + Zn (d)

Furthermore, quantitative analysis of treated charcoal was carried out. The analysis was carried out by the wet digestion method using aqua regia. The extracts were then measured for total Cu, Fe, and Zn levels using AAS. The results of the analysis are listed in Table 2, 3, and 4.

Table 2. Results of Activated Charcoal Analysis after Soaking in $CuSO_4$ Solution

Treatment (charcoal : solution (w/v))		Water content (%)	Ash content (%)	Total Cu (ppm)
B1	Control	1.59	6.86	96
	CuSO4 1N (1 : 3)	0.38	6.41	11,443
	CuSO4 1N (1 : 5)	0.85	6.45	8,749
	CuSO4 1N (1 : 7)	0,76	6.47	7,850
	CuSO4 2N (1 : 3)	1.75	6.43	9,002
	CuSO4 2N (1 : 5)	1.68	6.28	6,866
	CuSO4 2N (1 : 7)	1.12	6.38	6,109
B2	Control	2.14	5.67	104
	CuSO4 1N (1 : 3)	0.80	6.59	9,501
	CuSO4 1N (1 : 5)	0.93	6.80	8,530
	CuSO4 1N (1 : 7)	0.82	6.79	6,845
	CuSO4 2N (1 : 3)	1.61	6.91	8,541
	CuSO4 2N (1 : 5)	1.06	6.57	6,293
	CuSO4 2N (1 : 7)	1.65	6.32	5,951
T1	Control	0.51	2.04	36
	CuSO4 1N (1 : 3)	2.33	2.61	4,260
	CuSO4 1N (1 : 5)	521	2.41	3,206
	CuSO4 1N (1 : 7)	6.90	2.83	4,064
	CuSO4 2N (1 : 3)	0.33	3.01	4,880
	CuSO4 2N (1 : 5)	0.60	2.84	4,062
	CuSO4 2N (1 : 7)	1.26	2.45	3,509
T2	Control	4.49	3.22	91
	CuSO4 1N (1 : 3)	4.10	2.95	7,435
	CuSO4 1N (1 : 5)	0.85	3.01	7,741
	CuSO4 1N (1 : 7)	1.11	3.29	10,775
	CuSO4 2N (1 : 3)	0.36	3.32	3,558
	CuSO4 2N (1 : 5)	0.45	3.02	5,513
	CuSO4 2N (1 : 7)	0.41	3.17	5,968

Note: B1 = bamboo activated charcoal 600°C steamed 90 minutes; B2 = bamboo activated charcoal 700°C; steamed 90 minutes; T1 = coconut shell activated charcoal 600°C steamed 90 minutes; T2 = coconut shell activated charcoal 700°C steamed 90 minutes.

Based on the data in Table 2 it is known that in general, the total Cu content in B1 is higher than the Cu content in B2, while the Cu content in T2 is higher than the Cu content in T1. According to the data in Table 1, B2 adsorption on iodine is higher than B1, meaning B2 has more pores. However, because the bamboo activated charcoal is dominated by macropores, the Cu^{2+} which initially made it into the pore can be lost because it was washed when the charcoal was cleaned.

In coconut shell activated charcoal, the adsorption power of T2 is higher than T1, so that Cu is adsorbed more. The optimal combination of treatments for each raw material is immersion B1 in CuSO4 1N (1: 3), hereinafter referred to as C1 fertilizer and T2 immersion in CuSO4 1N (1: 7) which is hereinafter referred to as C2 fertilizer. In addition to Cu, in this experiment immersion was also carried out in $FeSO_4$ and $ZnSO_4$. The results of the analysis of activated charcoal soaked in $FeSO_4$ and $ZnSO_4$ are presented in Table 3 and Table 4.

Table 3. Results of Activated Charcoal Analysis after Soaking in $FeSO_4$ Solution

Treatment (charcoal : solution (w/v))		Water content (%)	Ash content (%)	Total Fe (ppm)
B1	Control	1.59	6.86	303
	FeSO4 1N (1 : 3)	0.94	6.13	2,034
	FeSO4 1N (1 : 5)	1.37	6.30	2,499
	FeSO4 1N (1 : 7)	1.06	6.16	3,797
	FeSO4 2N (1 : 3)	1.55	6.28	4,241
	FeSO4 2N (1 : 5)	2.31	5.93	5,476
	FeSO4 2N (1 : 7)	2.47	6.15	4,748
T2	Control	4.49	3.22	294
	FeSO4 1N (1 : 3)	1.68	2.54	1,266
	FeSO4 1N (1 : 5)	1.69	2.34	5,448
	FeSO4 1N (1 : 7)	1.85	2.40	4,775
	FeSO4 2N (1 : 3)	1.98	2.65	6,581
	FeSO4 2N (1 : 5)	1.80	3.20	7,611
	FeSO4 2N (1 : 7)	2.03	3.22	7,051

Note: B1 = bamboo activated charcoal 600°C steamed 90 minutes; T2 = coconut shell activated charcoal 700°C steamed 90 minutes

Based on the data in Table 3, the optimal treatment combination for each raw material is B1 immersion in $FeSO_4$ 2N (1: 5), hereinafter referred to as F1 fertilizer and soaking T2 in $FeSO_4$2N (1: 5) which is hereinafter referred to as F2 fertilizer. The amount of Fe adsorbed activated charcoal is relatively lower compared to Cu absorption, this is because the radius of the Fe atom is larger than the Cu atom.

Table 4. Results of Active Charcoal Analysis after Soaking in $ZnSO_4$ solution

Treatment (charcoal : solution (w/v)		Water content (%)	Ash content (%)	Total Zn (ppm)
B1	Control	1.59	6.86	23
	$ZnSO_4$ 1N (1 : 3)	1.97	6.00	6,603
	$ZnSO_4$ 1N (1 : 5)	1.74	5.94	4,648
	$ZnSO_4$ 1N (1 : 7)	1.85	6.18	3,879
	$ZnSO_4$ 2N (1 : 3)	2.36	6.09	4,290
	$ZnSO_4$ 2N (1 : 5)	1.87	6.24	2,873
	$ZnSO_4$ 2N (1 : 7)	2.14	6.15	2,856
T2	Control	4.49	3.22	8
	$ZnSO_4$ 1N (1 : 3)	1.96	2.60	6,343
	$ZnSO_4$ 1N (1 : 5)	1.79	2.81	4,443
	$ZnSO_4$ 1N (1 : 7)	1,60	2.96	2,638
	$ZnSO_4$ 2N (1 : 3)	1.78	3.07	4,441
	$ZnSO_4$ 2N (1 : 5)	1.93	2.60	2,205
	$ZnSO_4$ 2N (1 : 7)	1.75	2.77	1,900

Note: B1 = Bamboo activated charcoal 600°C steamed 90 minutes; T2 = coconut shell activated charcoal 700°C steamed 90 minutes

Based on the data in Table 4, the optimal treatment combination for each raw material is B1 immersion in $ZnSO_4$ 1N (1: 3) which is then referred to as Z1 fertilizer and T2 immersion in $ZnSO_4$ 1N (1: 3) which is then referred to as Z2 fertilizer. The amount of Zn which activated adsorbed charcoal is relatively lower compared to the absorption of Cu and Fe, this is because the radius of Zn atoms is smaller than Cu and Fe atoms, so even though Zn atoms can be adsorbed by activated charcoal, they can also be lost during the washing process.

3.3. Result of testing of nutrient release in products

This experiment was aimed to determine the release of Cu^{2+}, Fe^{2+}, and Zn^{2+} in the slow release fertilizer produced. The Experiment was carried out by extracting fertilizers with distilled water and 2% citric acid at extraction time 0, 15, 30, 45, and 60 minutes. The test results are presented in Table 5 - Table 10.

Table 5. Results of Cu Fertilizer Extraction with distilled water

Fertilizer	Cu content (ppm) in extraction time					Total Cu content (ppm)
	0'	15'	30'	45'	60'	
C1	61	67	67	85	104	11,443
C2	33	41	36	46	76	10,775

The data in Table 5 shows that Cu^{2+} has been dissolved in distilled water even without shaking. Distilled water extracted Cu^{2+} at C1 were higher than C2, so Cu^{2+} was more easily available in C1 fertilizer.

Table 6. Results of Fe Fertilizer Extraction with distilled water

Fertilizer	Fe content (ppm) in extraction time					Total Fe content (ppm)
	0'	15'	30'	45'	60'	
F1	0	1	1	1	7	5,476
F2	2	1	1	2	2	7,611

Based on the data in Table 6, extracted Fe^{2+} content are very low. This is because during the drying process, Fe^{2+} is oxidized to Fe^{3+} which is more stable and difficult to dissolve in distilled water.

The results of extraction of fertilizers with distilled water indicate the amount of elements available on the soil with neutral conditions that can be immediately adsorbed by plants. The distilled water content of Cu^{2+}, Fe^{2+},and Zn^{2+} is much lower than the total elements in activated charcoal after immersion, this data shows that the release of nutrients occurs slowly.

Table 7. Results of Zn Fertilizer Extraction with distilled water

Fertilizer	Zn content (ppm) in extraction time					Total Zn content (ppm)
	0'	15'	30'	45'	60'	
Z1	36	36	45	50	50	6,603
Z2	45	63	55	52	74	6,343

Table 8. Results of Cu Fertilizer Extraction with 2% citric acid

Fertilizer	Cu content (ppm) in extraction time					Total Cu content (ppm)
	0'	15'	30'	45'	60'	
C1	7,834	8,985	9,080	9,656	10,176	11,443
C2	6,219	8,418	8,138	8,513	8,692	10,775

Citric acid has the ability to chelate metal ions and keep them in solution in pH conditions where they should settle. The extraction results with citric acid indicate that the Cu^{2+} will be available entirely after being shaken for 60 minutes.

Table 9. Results of Fe Fertilizer Extraction with 2% citric acid

Fertilizer	Fe content (ppm) in extraction time					Total Fe content (ppm)
	0'	15'	30'	45'	60'	
F1	1,220	2,121	2,176	2,230	2,548	5,476
F2	1,653	2,283	2.476	2,476	2,450	7,611

Citric acid forms a more stable chelate with Fe^{3+}. However, the amount of Fe extracted with citric acid tends to be much lower than the total Fe in fertilizer.

Table 10. Results of Zn Fertilizer Extraction with 2% citric acid

Fertilizer	Zn content (ppm) in extraction time					Total Zn content
	0'	15'	30'	45'	60'	(ppm)
Z1	5,699	6,058	6,195	6,162	6,330	6,603
Z2	5,964	5,911	5,960	6,221	6,483	6,343

The results of Zn fertilizer extract with citric acid showed that without any shaking Zn^{2+} extracted approached the total Zn level in fertilizer. The results of this test showed that in the fertilizer made in this study, Fe was adsorbed more strongly by activated charcoal compared to Cu and Zn. Some of the properties that affect activated charcoal adsorption are the chemical-physical properties of adsorbents such as pore size, and chemical composition of activated charcoal, chemical-physical properties of adsorbates such as molecular size and polarity, liquid phase properties such as pH and temperature and length of the adsorption process.

Further tests were also carried out to determine whether or not the nutrients in activated charcoal were lost due to washing. The test was carried out by washing 25x coconut shell activated charcoal + Cu (C2), then the fertilizer was dried and observed with EDX (Figure 6). From Figure 6 it is known that Cu elements are still found in activated charcoal even though it has been washed as much as 25x. This illustrates that Cu is strongly adsorbed in activated charcoal and is not easily released.

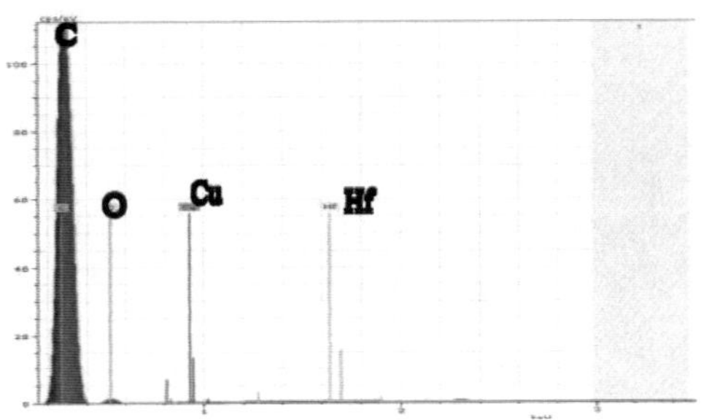

Figure 5. Observation results of EDX on C2 washed 25x

4. Conclusions

According to the results of the study, bamboo and coconut shell activated charcoal made by the activation process at temperatures of 600°C and 700°C and steamed for 90 minutes fulfills SNI 06-3730-1995, except for adsorption capacity on iodine and methylene blue. Only coconut shell activated charcoal is activated at a temperature of 700°C and steamed for 90 minutes which has adsorption capacity on iodine according to SNI 06-3730-1995. This shows that the activation process carried out is not enough to open the charcoal pores. However, activated charcoal can be used to adsorb added micronutrients. The adsorption capacity of activated charcoal on nutrients added is different, with the total adsorbed Cu content higher than total Fe and Zn. This is influenced by the pore size of each activated charcoal and the size of Cu, Fe and Zn atoms.

Based on the testing of nutrient release rate, it is known that the fertilizer made in this study is slowly released, with a different release rate, ie Zn is more easily released compared to Cu and Fe. In addition, it is also known that the elements in fertilizer are not easily lost due to washing experiment because they are adsorbed strongly enough by activated charcoal.

References

[1] [UNIDO] United Nations Industrial Development Organization [IFDC] International Fertilizer Development Center. 1998. Fertilizer Manual. Dordrecht: Kluwer Academic Publishers.

[2] Alfianto R. 2011. Kajian pembuatan arang aktif dari sekam padi dengan teknik pelarutan silika [skripsi]. Bogor: Fakultas Pertanian. Institut Pertanian Bogor.

[3] Sudrajat R, Soleh, S. 1994. *Petunjuk Teknis Pembuatan Arang Aktif.* Bogor: Bagian Proyek Litbang Pemanfaatan Hasil HTI Pusat Litbang Hasil Hutan dan Sosial Ekonomi Kehutanan Badan Penelitian dan Pengembangan Kehutanan.

[4] Nurhayati T. 2000. Percobaan pembuatan arang aktif dari bambu. *Prosiding Sarasehan Penelitian Bambu Indonesia.* Serpong 21-22 Juni 1994. Hlm. 177-184.

[5] Sudrajat R, Pari G. 2011. *Arang Aktif Teknologi Pengolahan dan Masa Depannya.* Jakarta: Balai Penelitian dan Pengembangan Kehutanan – Kementrian Kehutanan.

[6] [BSN] Badan Standardisasi Nasional. 1995. *Arang Aktif Teknis*. Jakarta: BSN; (SNI 06-3730-1995)

The effect of *Curcuma zedoaria* (Berg) Rosc. plant rhizome extracts on proliferation and viability of myeloid and fibroblas cancer cells

Saefudin

Biology Research Centre, Indonesian Institute of Science, Cibinong Science Center, Bogor, Indonesia.

email: saefudinkahfi@gmail.com

Abstract. This research aimed to look into the effect of extracts from *Curcuma zedoaria* (Berg) Rosc. plant rhizomes on the proliferation of K-562's myeloid cancer and WeHi-164's fibroblast tumor cells. Two types of extracts, comprising water-media and ethanol-media extracts were prepared; and then both tested at varying concentrations (0, 0.1, 1.0, 10.0, and 100.0 g/ml). Proliferation activities were performed on all extract concentrations, either water or ethanol media. Observation focused on morphology-anatomy characteristics of cancer cells. Both extract types (water and ethanol media) inflicted anti-proliferation actions on cancer cells. The lowering/decreasing intensity of the cell proliferation as affected by plant extracts was significant at high extract concentration (100.9 g/ml), with water as well as ethanol media; but less effective at low concentrations (0.1-10.0 g/ml). Anti- proliferation activities by plant extracts at low concentration altered morphology, shapes, and sizes of K-562's and WeHi-164's cells, but their cell wall remained intact (unchanged). The use of plant extracts with ethanol media at high concentration altered the shapes/morphology, sizes of K-562's myeloid cells, fibroblast cells enlarging, cell wall fractured, and the cells fragmented. Plant extracts with ethanol media afforded stronger anti-proliferation activities than those with water media.

1. Introduction

Curcuma zedoaria (Berg) Rosc. (temu putih) plants belong to the member of Zingiberaceae family. The origin of *C. zedoaria* plants is not precisely known, but presumably they are indigenous from North West India. The bio-prospects of *C. zedoaria* plants could be as medicine/drug plants and cooking aroma. Currently, those plants have been cultivated under the tree stands for production forest of consecutively teak (*Tectona grandis* Lf), tusam (*Pinus merkusii* Jungh at de Vriese), and mahoni (*Swietenia macrophylla*) species. The rhizomes of *C. zedoaria* plants contain essential compounds for traditional remedies and drug industries [1].

Curcuminoide compounds such as curcumine, di-methoxy-curcumine, and bis-di-methoxy-curcumine typify as bioactive components in *Curcuma* genus, which have been known able to inflict citotoxic effects on OVCAR-3 (human ovarian cancer cells) and have been traditionally used as a remedy for cancers at vagina labium [2]. Curcumin, the phenolic pigment, appears as deep yellow or orange in colour, the substance that exists in the rhizomes of many species of *Curcuma* family. According to [3], the amount of curcumin in rhizomes of *C. zedoaria* by HPTLc analysis is 16.715 mg/g. Curcumenone and curcumenol are two of these compounds of *C. zedoaria* that have anti-proliferative effects to induce apoptosis in MCF-7 cells by inhibiting cancer cell proliferation [4]. Petroleum ether and chloroform root extracts of plant *C. zedoaria* possess significant anti-inflammatory activity, which was tested in rats [5].

The research results of [6], ermanin and curcumin were the most effective among the isolated compounds from *C. zedoaria* rhizomes, with inhibition of 67.58±3.82 and 95.36±1.58 g/mL, respectively, against collagen induced platelet aggregation at 100 M. Platelet-rich plasma has is its ability to stimulate bone and soft tissue healing. The presence of ermanine and curcumin in the C. zedoria rhizome is possible for the treatment/prevention of cardiovascular diseases.

Curcumine typifies as polyphenol compound that could exert biology activities as antioxidant, anti-mutagenic, anticoagulant, antifertility, anti-diabetic, antibacterial, antifungi, antiprotooa, antivirus, and anti-fibrosis agents [7]. Other researchers reported that curcumine could inhibit the proliferation of carcinomia heptatocellular cells in the human body by inducing (triggering) the stresses on endoplasmic reticulum and dysfunction on mitochondria [6]. In India, China, and South East Asia countries such as Indonesia, people there utilize yellow-colored substances from curcuma plants as additional food, ingredients as well as drugs, which so far do not inflict toxic effects [8].

New sesquitepene compounds of the so-called furanogermenone that were isolated from the rhizomes of *C. zedoaria* brought about the effect on increasing the SGOT and SGPT in mice, which were previously inducted (triggered) by CCl_4; and did not cause serious damages to their liver [9]. Ethyl-p- metoxycimanic which was present in their rhizomes of *C. zedoaria* also revealed its antifungal characteristics [10]. B-1 fraction that was obtained from the purification of B fraction was inherently a single compound with the predicted formula of $C_{15}H_{16}O_2$ typified as 1(10),5,7,(11) 8, Guaiantetraen-12,8-olide (gweiikurkulakton) compound, which belonged to anti-proliferation bioactive compound further proved efficacious against HeLa cells [11]. Anti-proliferation activities performed by *C.zedoaria* extracts at high concentration (10%) could effectively inhibit the growth/proliferation of HeLa cells, and reach LC_{50} value at the extract concentration equal to 60.3 μg/ml [12].

Relevantly in these research activities, it would be examined the rhizome extracts of *Curcuma zedoaria* plants, which grew under different environments. Further, the rhizome extraction was performed using two different solvent liquids, which comprised water and ethanol (95%) media. Meanwhile, anti-proliferation activities as expectedly afforded by *C. zedoaria* rhizome extracts and their cytotoxic effects were tested on two different cancer cells, which covered K-562 (myeloid cancer cells) and WeHI-164 (fibroblast tumor cells).

2. Materials and methods

2.1. Materials

Main research materials were the rhizomes of *C. zedoaria* plants, collected from the Forestry District Division (BKPH) of Majenang, West Banyumas (Central Java). Further, the rhizomes of the same origins were partly cultivated at the Cibinong Sciense Centre's experiment garden in Cibinong, Bogor (West Java). The *C. zedoaria* rhizomes were allowed for some time to dry naturally under the roof, then milled/ground to pieces, and sieved (using 8-mesh screen) such that the rhizomes in flour shape were obtained. *C. zedoaria* rhizome flour was then extracted using two kinds of liquid media, i.e. ethanol and water solvent. Extracts from *Echinacea purpurea* plants were used as a positive control. In this research, two kinds of cancer cells which were tested, comprised K-562 (myeloid cancer cells) and WeHi 164 (fibroblast tumor cells), simply as a collection of the Animal Husbandry Faculty, Bogor Agricultural University, Bogor-Indonesia.

2.2. Activities of anti-proliferation agent

Anti-proliferation could be defined as the effective action of anti-proliferation agents at their specific concentration (percentage) that could inhibit the growth/proliferation of particular cells as much or greater than 50% (LC_{50}) as their normal growth (proliferation) capacity. The cells which were tested in vitro comprised K-562 (CML myeloid cancer cells) and WeHi-164 (sarcoma-fibroblast tumor cells). Those two kinds of line cells (cancer cells) were each cultured at the concentration equal to 10^3 cells/ml in the so-called 24-well dish using the growth media of DMEM (Dulbecco's modified eagle medium) and 10 % FCS/Foetal calf serum [13]. The culture media that contained the two kinds of cultured cancer cells (K-562 and WeHI-164) were then placed into the CO_2 incubator for 24 hours.

The treatment in this regard was the *Curcuma zedoaria* rhizome extracts with two kinds of solvent liquid (ethanol and water media), which supposedly could act as anti-proliferation agent against those cancer cells. Further, the rhizome extracts (with either ethanol or water media) were made varying in their concentration (i.e. 0, 0.1, 1.0, 10.0, and 100.0 g/ml; also regarded as the treatment), and then each was put (infected) into the growth media that already contained the cultivated cancer cells (K-562 and WeHi-164). Each of the extract concentrations was replicated three times. The incubation for the infected cultured cancer cells, after the first 24-hour duration, was continued again for another 24 hours. Afterwards, the tested/infected cancer cells could be harvested; and were further counted of their total cells in the dishes using the so-called hemocytometer instrument with the aid of trypan blue's coloring agent. Results of the cell counting were compared with those (results of the cell counting) for negative control and positive control (*Echinacea purpurea* plant extracts).

2.3. Toxicity test and cell morphology

Sediment of the infected cultured cancer cells that occurred due to the treatment (infection) by supposedly anti-proliferation agent (*C. zedoaria* rhizome extracts) was immersed in glutaraldehyde (2%) liquid for 24 hours, then added with buffer chocodylate, and allowed to stand in the immersion, until the sediment dissolved to form a solution. The solution was further centrifuged vigorously, such that new sediment was formed; and afterwards the supernatant (clear liquid part) was separated (discarded). The remaining sediment was then added with osmium tetraoxyde (1%) liquid until it was immersed completely and then kept there for one hour. Afterwards, the sediment was separated from the immersing liquid and, then dried, and further immersed in alcohol (80%) for 20 minutes. After the immersion, the sediment was separated and then suspended by adding butanol liquid; and the sediment suspension was placed on the cover slip, which had already been stuck to the aluminum stub. In this way, the suspension would dry gradually, leaving behind the dried sediment. The dry sediment on the cover slip further underwent the vacuuming for 20 minutes, and then meticulously observed using Scanning Electron Microscope (SEM 5000) device for examining and scrutinizing the morphology of the two tested cancer cells (K-562 cells, and WeHi-164 cells). Results of the SEM's morphology scrutiny on those two kinds of cells were compared with results of the SEM scrutiny on the Hella cells, already treated with anti-proliferation agent (*C. zedoaria* plant extracts) obtained from the previous research [12].

3. Results and discussion

Viability (i.e. living cells which are still able to proliferate for their growth) of the tested cancer cells in the initial testing averaged about 95.3% of their normal capable growth (pointed-out in the negative control) (Table 1).

Table 1. Average inhibition of the tested cancer cells/IC50 of *C. zedoaria* rhizome extracts, at varying extract concentrations; and with two types of extracts.

Treatment	Average cell viability (%) / IC50			
	K -			K +
Extract concentrations		Low	High	
	95.3a	72.4a	39.7b	24.6b
Types of extracts (solvent media)		Water solvent	Etanol Solvent	
	95.3a	43.4ab	25.5b	24.6b

Remarks: Average figures followed horizontally by the same letters are not significantly different (based on Duncan's multiple range test at 5% level; a > b > c); cell inhibition (%) = 100% - cell viability (%); Low = 0.1-10 g/ml; High = 100 g/ml; K - = negative control; K + = positive control (extracts from *Echinacea purpurea* plants)

Testing results on anti-proliferation activities revealed that the treatment by suspected anti-proliferation agent (*C. zedoaria* rhizome extracts) at their low concentration (0.1-10 µg/ml) did not inflict significant effect on the proliferation capability of K-52's as well as WEHI-164's cancer cells, either using water solvent or using ethanol solvent as the extraction media (Table 1). Increasing the concentration of *C. zedoaria* rhizome extracts (to 100 µg/ml) brought about more intensive inhibition effects or inflicted more active anti-proliferation actions significantly on the tested cancer cells (K-52 and WeHi-164), compared to the situation (inhibition effects/anti-proliferation actions) in the negative control. It was indicated that activities of anti-proliferation by rhizome extracts could strongly relate to the chemical content of their extractives.

Cell viability at low concentration of *C. zedoaria* rhizome extracts (0.1-10 µg/ml) reached 40.3-92.0% (in range) for either of the two kinds of cancer cells (K-562 and WeHi-164), or 72.4% as average. That viability was regarded as still high, indicating that the anti-proliferation activities by low concentration *C. zedoaria* extracts were less active in inhibiting the cancer cell proliferation (less inhibition effect). However, high concentration of *C. zedoaria* extracts (100 µg/ml) apparently exhibited the most actively inhibiting the proliferation of the tested cancer cells (K-562 and WeHi-164), as indicated by their low cell viability (22.0–47.3% in range or 39.7% in average).

Such cancer cell viability (39.7% in average) at high concentration of *C. zedoaria* extracts (100 µg/ml) was significantly lower than the viability (72.4% in average) at low extract concentration (0.1-10 µg/ml); and also lower than the viability of the negative control (95.3% in average) (Table 1), whereas the difference in viability between low concentration *C. zedoaria* extracts (72.4% in average) and the negative control (95.3% in average) was not significant. Concurrently, the difference in viability between high concentration *C. zedoaria* extracts (39.7% in average) and the positive control (24.6% in average) was not significant either (Table 1). Judging from those quantitative results, accordingly, high concentration of *C. zedoaria* rhizome extracts (100 µg/ml) associated with the average cancer-cell viability (39.7%) could be regarded as their reliable LC_{50} value. This is because the cell inhibition achieved by those extracts could reach ≥50% (inhibition: 100%–39.7% = 60.3%); and that IC_{50} value applied to both K-562 and WeHi-164 cancer cells.

Further, Table 1 revealed that *C. zedoaria* rhizome extracts at low concentrations (0.1–10 µg /ml) as described before still revealed their anti-proliferation activities, whereby unfortunately the viability or proliferation rate of the tested cancer cell was still greater than or > 50% (72.4%). In this situation, therefore, the capability of *C. zedoaria* extracts at 0.1–10 µg /ml concentrations to inhibit the cancer cell proliferation was still less than or < 50% (inhibition: 100%–72.4% = 27.6%). On the other hand, *C. zedoaria* rhizome extracts at high concentrations/dosages (100 µg /ml) also as described before could reduce the cancer-cell viability to 37.9% (viability < 50%); or concurrently could inhibit the cancer cell proliferation as much as >50% (inhibition: 100%–39.7% = 60.3%). Accordingly, again such high dosages (100 µg /ml concentrations) was confirmed as the reliable LC_{50} value for *C. zedoaria* rhizome extracts.

The main compounds in *C. zedoaria* extracts, such as particularly arylheptanoid (curcuminoid), volatile (essential) oils with their varying monoterpenes and sesquiterpenes, and polysaccharides could exhibit remarkable anticancer actions; and therefore were immensely effective to press down the growth/proliferation of K-562's myeloid cells and WeHi-164's tumor cells; or were so to other cells that might exhibit cancer behavior [11,14]. The increase in cell proliferation could occur due to the activation of several oncogenes involved in the mitogenic signals such as specific race, whereas the inhibition action on biology processes which were deadly to the cancer cells was brought about by the so-called bcl2's over-expression [14].

Use of different solvent liquids (water and ethanol media) for the extraction of *C. zedoaria* rhizome apparently brought about prominently different extract performances as antiproliferation agents against the cancer cells (Table 1). It revealed that *C. zedoaria* extracts with water media at particular concentrations brought specific inhibition effect on the K-562's cancer cells (viability/proliferation: 41.0–90.3%) and WeHi's cancer cells (viability/proliferation: 42.0-92.0%). However, the extracts with ethanol media, at similar concentrations (as water media) exerted stronger inhibition effects on the cancer cells (indicated by lower cell viability/proliferation rate) consecutively for K-562's cancer cells (viability/proliferation: 22.0–77.6%) and for WeHi's cancer cells (viability/proliferation: 22.6–85.3%). Results of Duncan's multiple range test confirmed that allegation (Table 1), whereby the average cell viability/proliferation rate was 43.4% (inhibition: 100–43.4% = 56.6%) due to the use of *C. zedoaria* extracts with water media. On the other hand, due to the use of *C. zedoaria* extracts with ethanol media, the average cell viability/proliferation rate of cancer cells was significantly lower (cell viability lower [25.5%]; or greater inhibition [100% - 25.5% = 74.5%]).

Further still related, in inhibiting the cancer cells the performance of *C. zedoaria* rhizomes' ethanol extracts (cell viability 25.5% or cell inhibition 74.5%) was statistically comparable with that of positive control/K+ (cell viability 24.6% or cell inhibition 75.4.5%) (Table 1). Meanwhile, in doing so the performance of *C. zedoaria* rhizomes' water extracts (cell viability 43.4% or cell inhibition 56.6%) as stated before was statistically lower than that of *C. zedoaria* rhizomes' ethanol extracts, but still better than that of negative control/ K- (cell viability 95.3% or cell inhibition 4.7%). Results of the previous research which used *C. zedoaria* extracts (with ethanol media as well) could inhibit the proliferation of the cervix cancer cells in the range of 32.0-87.6% (inhibition); or cell viability range of 68.0–12.4% [12]. In that research, the average cell viability reached 95.3% (under normal condition or in negative control), whereas concurrently the related LC_{50} value for *Echinacea* sp. extracts was achieved at 24.5% cell-viability/proliferation (or cell inhibition: 100%–24.5% = 75.5% [>50%]). Compared to the LC_{50} value in this research results (cell viability = 39.7%; or cell inhibition = 60.3%), then in inhibiting the cancer cells, the performance of *C. zedoaria* rhizome extracts was slightly lower than that of *Echinacea* sp. extracts.

Scrutiny using the scanning electron microscopy (SEM) could provide clear illustration associated with the effect of *C. zedoaria* rhizome extracts (with ethanol as well as water media) at high concentration (100 g/ml) on the visual appearance of the tested cancer cells (K-562 and WeHi-184) (Figure 1). Toxicity of *C. zedoaria* rhizome extracts with water media affected the shape (morphology) of K-562's cancer cells which became enlarged, while the shape/morphology of WeHi-164's cancer cells was unchanged; both kinds of cells were compared to their corresponding cells in the control. Meanwhile, the toxicity of *C. zedoaria* rhizome extracts with ethanol media revealed that the line cells tended to get bigger, cell wall destroyed, cell fragmentation (necrosis) occurred. Therefore, this visual illustration confirmed that *C. zedoaria* rhizomes' ethanol extracts inflicted stronger antiproliferation effects on the tested cancer cells than *C. zedoaria* rhizomes' water extracts, both at high concentration (100 μg/ml). At high extract concentration, the ethanol fraction was able to explore bioactive components in *C. zedoaria* rhizomes more perfectly than the water fraction. Ethanol extracts afforded better polarity than water extracts; and consequently, chemical compounds in *C. zedoaria* rhizome extracts such as flavonoids in free form as well as in their glycoside forms could be extracted more intensively.

According to [2], curcuminoide compounds in *Curcuma* genus were known to have cytotoxic effects on OVCAR-3 (human ovarian cancer cells); and traditionally already used a remedy for cancers at vagina labia. *C. zedoaria* extracts contained as much 0.1% as curcumine compounds [15]. Use of SEM could provide visual illustration about morphology changes (Figure 1) in the two kinds of tested cancer cells (K-562 and WeHi-164) at high concentration of *C. zedoaria* rhizome extracts (100 μg /ml).

From the viability data (Table 1), then it could be formulated that *C. zedoaria* rhizomes' ethanol extracts inflicted stronger antiproliferation actions on the cancer cells than *C. zedoaria* rhizomes' water extracts. Results of the SEM's scrutiny (Figure 1) provided stronger and clearer illustration regarding the effect of treatment (*C. zedoaria* rhizomes extracts with ethanol as well as water media) on the line cells (tested cancer cells). It appeared as visually disclosed before that the use of *C. zedoaria* rhizomes' water extracts induced K-562's cells to grow bigger, whereas WeHi-164's cells remained unchanged, compared to the control. On the other hand, the effects of using *C. zedoaria* rhizomes' ethanol extracts revealed that all line cells tended to grow bigger, cell wall destroyed to pieces (fractured), and all cells fragmented. Further, ethanol extracts at high concentration brought about stronger inhibition activities or enhanced more active anti-proliferation effects on the two kinds of tested cancer cells.

IOP Conf. Series: Earth and Environmental Science **359** (2019) 012010 doi:10.1088/1755-1315/359/1/012010

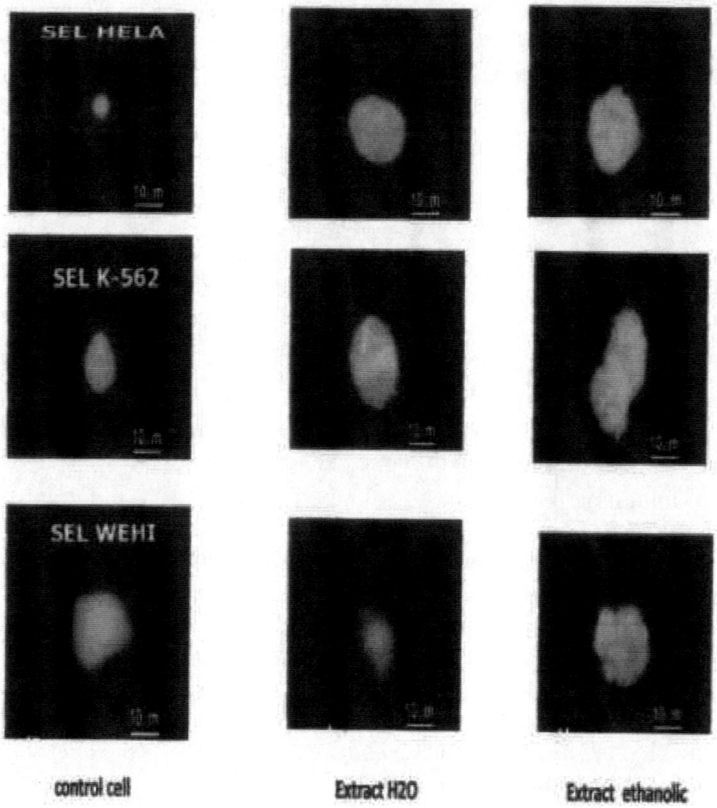

A **B** **C**

Figure 1. Scanning electron micrograph visually featuring the effect of *C. zedoaria* rhizomes extracts using water and ethanol media on the morphology of tested cancer cells (using SEM under 15,000 times magnification)); A. Control; B. Water-media extracts; C. ethanol-media extracts.

4. Conclussion and suggestions

The use of *C. zedoaria* rhizome extracts (with ethanol as well as water media) at low concentrations (0.1- 10 g/ml), as allegedly antiproliferation agent, seemed less effective in inhibiting the growth (proliferation) of the tested cancer cells (K-562 and WeHi-164). However, at high extract concentration (100 g/ml), there occurred notable changes in the cell morphology and sizes which became enlarged, cell wall fractured, and all cells fragmented.

At high extract concentration, antiproliferation activities became more pronounced; and the extract effect on inhibiting the proliferation of cancer cells tended to be more severe or brought about the cell viability less than 50% (39.7%) or inhibition of cell growth greater than 50% (100%–39.7% = 60.3%). Therefore, extract concentration at 100 g/ml at which the cell viability was 39.7% (or cell growth/proliferation inhibition 60.3%) was regarded as the reliable LC_{50} value for the *C. zedoaria* extracts (as the alleged antiproliferation agent).

C. zedoaria extracts, either with ethanol or water media, exhibit potentiality as antiproliferation agent against K-562's and WeHi-164's cancer cells. However, *C. zedoaria* extracts with ethanol media inflicted stronger antiproliferation actions compared to the extract with water media. At high concentration (particularly with ethanol media) strongly deserves thoroughly further research continuation.

Acknowledgements
The author thanks to Prof. Dr. Chairul Apt, who has provided guidance and inputs at this paper.

References

[1] Lobo R, Shirwaikar A, Prabhu KS and Shirwaikar A. *Curcuma zedoaria* Rosc, (white turmeric): A review of its chemical, pharmacological and ethnomedicinal properties. *Journal of Pharmacy and Pharmacology*. 2009;61(1):13–21.
[2] Syu WJ, Shen C, Don MJ, OU JC, Lee GH and Sun C. Cytotoxicity of curcuminoids and some novel compounds from Curcuma zedoaria. *Journ Nat Prod*. 1998;61(12):1531–4.
[3] Lavania SRS. Histochemical localization of curcumin and its significance in chemotypic characterization of selected species of Curcuma L. *Industrial Crops and Products*. 2015;65:175–179.
[4] Hamdi OAA, Ye L, Kamarudin MNA, Hazni H, Paydar M, et al. Neuroprotective and antioxidant constituents from *Curcuma zedoaria* rhizomes. Original Article. *Rec. Nat. Prod.* 2015;9(3):349–355.
[5] Kausik ML and Jalalpure SS. Anti inflammatory efficacy of *Curcuma zedoaria* Rosc. root extracts. Research Article. *Asian Jour. Pharm. Clinical Research*. 2011;4(3):90–92.
[6] Chen JJ, Tsai TH, Liao HR, Chen LC, Kuo YH, Sung PJ, et al. New sesquiterpenoids and anti-platelet aggregation constituents from the rhizomes of *Curcuma zedoaria*. *Molecules Journ.* 2016;21(10):1385.
[7] Choi H, Lim JE and Hong JH. Curcumin interrupts the interaction between the androgen receptor and Wnt/β-catenin signaling pathway in LNCaP prostate cancer cells. *Prostate Cancer Prostatic Dis.* 2010;13(4):343–349.
[8] Mazzarino L, Silva L, Curta J, Licínio MA, Costa A, Pacheco LK, et al. Curcumin-loaded lipid and polymeric nanocapsules stabilized by nonionic surfactants: an in vitro and in vivo antitumor activity on B16-F10 melanoma and macrophage uptake comparative study. *Journ Biomed Nanotechnol*. 2011;7(3):406–414.
[9] Xie CY, Yang W, Ying J, Ni Q, Pan X, Dong JH, et al. B-cell lymphoma-2 over-expression protects δ-elemene-induced apoptosis in human lung carcinoma mucoepidermoid cells via a nuclear factor kappa B-related pathway. *Biol. Pharm. Bull.* 2011;34(8):1279–1286.
[10] Prakobwong S, Gupta SC, Kim JH, Sung B, Pinlaor P, Hiraku Y, et al. Curcumin suppresses proliferation and induces apoptosis in human biliary cancer cells through modulation of multiple cell signaling pathways. *Carcinogenesis*. 2011;32(9):1372–1380.
[11] Ros S. *Karakterisasi Kimiawi dan Aktivitas Antiproliferasi Sel Lestari Tumor Serta Aktivitas Fagositosis Secara In-vitro dari Fraksi Bioaktif Rimpang Temu Putih (Curcuma zedoaria Rosc.)*. Pascasarjana IPB, Bogor.; 2006.
[12] Saefudin S, Syarif F and Chairul C. Potensi antioksidan dan aktifitas antriproliferasi ekstrak kunyit putih (*Curcuma zedoaris* (Rosc.) pada sel hela. *Widyariset*. 2014;17(3):381–389.
[13] Priosoeryanto BP, Tateyama S, Yamaguchi R and Uchida K. Establishment of a cell line (MCM-B2) from a benign mixed tumour of canine mammary gland. *Research Veterinary Science.* 1995;58(3):272–276.
[14] Kitamura C, Nagoe T, Prana MS, Agusta A, Ohashi K and & Shibuya H. Comparison of *Curcuma* sp in Yakhushima with *C. aeruginosa* and *C. zedoaria* in Java by trn gene sequence, RAPD pattern, and essential oil component. *Jurnal of Natural of Medicines*. 2007;61(3):239–243.
[15] Ros S and Djamil R. Kadar kurkumin dan potensi antioksidan ekstrak etanol *Curcoma zedoaria* (Berg) Rosce, *Curcuma mangga* Val et Zyp, dan *Curcuma xanthoriza* Roxb. *Jurnal Kefarmasian Indonesia*. 2006;11(2):21–27.

IOP Conf. Series: Earth and Environmental Science **359** (2019) 012011 doi:10.1088/1755-1315/359/1/012011

Moisture-content based on drying schedule of seven Indonesian woods species

E Basri* **and R Pari**

Forest Products Research and Development Center
Research Development and Innovation Agency
Jl. Gunung Batu 5 Bogor, Jawa Barat, Indonesia

*email: denvig@yahoo.com

Abstract. A moisture-content based on drying schedule provides general guidance in determining the changes in temperature and humidity which is based on the average moisture content level of the wood being dried. The research aims to obtain a basic drying schedule for seven wood species based on their drying characteristics. The research is initiated by drying all these species at a temperature of 100 °C. The criterion used is the worst defect occurred in each wood such as end and surface, honeycomb defects, and deformation on the width of the wood. The result shows that each wood species has a different response to high temperature drying. The best drying properties is observed in Red Meranti wood (*Parashorea smythiesii*) and Jaha wood (*Terminalia arborea*). In contrary, Kiacret (*Spathodea campanulata*) and Kipasang (*Prunus javanica*) have the worst drying properties. Based on their drying properties, all of the species can be grouped into 4 drying schedules.

1. Introduction

Drying is a critical part of most wood products manufacturing process. The methods used and proper control are key to achieving the appropriate production level, quality, and costs. Wood products that will be exported to mainly four-season destination countries or used in room air-conditioned environment with very low relative humidity require moisture content about 10% or below. Therefore, the timber must be dried properly according to the required standard. Moisture-content based on drying schedule is general guidelines that include measuring of temperature and relative humidity changes based on the average moisture content of the wood. Based on the physical and drying properties of each wood, dry and wet-bulb temperatures are determined according to the level of certain moisture content [1]. Not every wood species is easy to be dried without developing any defect, because each wood species has its own sensitivity to drying temperature. Threfore the importance information about the drying properties of wood species, especially their relationship to the basic properties, such as the physical properties of the wood. Lack of understanding the drying schedule of each species leads to a high drying cost, drying time, and low quality of the finished products [2–5].

The research aims to obtain the basic drying schedules for seven wood species (i.e. Jaha, Kiacret, Kipasang, Red Meranti, White Meranti, Bira-Bira, and Bipa) based on their drying characteristics. The basic drying schedules of these species are still limited. A proper drying schedule of each species is an important factor in determining the success of the drying process. Commonly, the schedule used by industries is based on the change in the moisture content of the wood.

2. Materials and method

Seven wood species were used in this drying investigation, i.e. Jaha, Kiacret, Kipasang, Red Meranti, White Meranti, Bira-Bira, and Bipa. Information about each species is presented in Table 1.

Table 1. The seven wood species investigated

Local name	Botanical name	Wood origin
Jaha	*Terminalia arborea* K. et. V.	West Java
Kiacret	*Spathodea campanulata* P.B.	West Java
Kipasang	*Prunus javanica* Miq.	West Java
Red Meranti	*Parashorea smythiesii* Wyatt.Sm ex P.S. Ashton F.Heim.	East Kalimantan
White Meranti	*Parashorea tomentella* (Sym.) Meijer	East Kalimantan
Bira-Bira	*Barringtonia asiatica* (L.) Kurz.	Riau
Bipa	*Pterygota horsfieldii*	Papua

Physical properties, namely initial moisture content in green condition and wood shrinkages from green to air dry were measured according to ASTM D143-94 [6]. Total samples of each species for physical properties tests were 10 pieces (ten replications).The method for drying properties testing in this experiment was a quick-drying test (high drying temperature), which was a modification of the Terazawa Method [7]. Ten defect-free wood samples (2.5 cm × 10 cm × 20 cm) were taken from the green flat-sawn lumber for each species and weighed. The samples were then dried in an oven at 100 °C to constant weight is reached at an average moisture content of ± 2%.

A set of criteria were used to assess three types of defect. Size and number of defects taking place on the surface of the dried lumber were scored using a scaled set of values which ranged from 1 to 7 for initial checks as well as deformation, and from 1 to 6 for honeycomb defect. The lower of the values means the less of the defects while the higher of the values, the more severe the defects. The data for the drying properties from quick-drying test and condition of each sample was used to determine the optimum drying schedule for the wood.

By referring to the drying conditions (Table 2), dry-bulb temperature (DBT) and wet-bulb depression (WBD) (Table 3 and 4) as described in the Forest Products Laboratory manual [1], basic drying schedule of each species was determined.

Table 2. Drying condition based on degree of defects[a]

Types of defects	Drying conditions (°C)	Defect classes						
		I	II	III	IV	V	VI	VII
I. End and surface checks	Initial temperature	70	65	60	55	50	40	38
	Wet bulb depression	6.5	5.5	4	4	3	2	2
	Final temperature	95	90	85	80	70	65	50
II. Deformation	Initial temperature	70	66	58	54	50	40	38
	Wet bulb depression	6.5	6	5	4	4	3	3
	Final temperature	95	88	83	80	70	65	50
III. Honeycombing	Initial temperature	70	55	50	48	48	40	-
	Wet bulb depression	6.5	4.5	4	3	3	2.5	-
	Final temperature	95	83	77	73	71	70	-

Remarks: [a]Modifiied from Terazawa Method [7]. 1 to 7 are its category. 1 = <5%, 2 = 5-10%, 3 = 10.1-20%, 4 = 20.1-40%, 5 = 40.1-60%, 6 = 60.1-80%, 7 >80%.

Table 3. Drying schedules as developed based on dry-bulb temperature (DBT).

Moisture content at start of step (%)	Dry bulb temperature (DBT) (°C)													
	T1	T2	T3	T4	T5	T6	T7	T8	T9	T10	T11	T12	T13	T14
Initial ~ 30	38	40	45	45	50	50	55	55	60	60	65	70	75	80
30 ~ 25	42	45	50	50	55	55	60	60	65	65	70	75	80	90
05 ~ 20	42	50	55	55	60	60	65	65	70	70	70	75	80	90
20 ~ 15	45	55	60	60	65	65	70	70	70	75	80	80	90	95
< 15	50	65	70	80	70	80	70	80	70	80	80	80	90	95

Source: (1)

Table 4. Patternized drying schedules based on wet-bulb depression (WBD).

Initial MC and range (%)						Wet-bulb depression (WBD) (°C)							
A	B	C	D	E	F								
40	50	60	75	90	110	1	2	3	4	5	6	7	8
40 ~ 30	50 ~ 35	60 ~ 40	75 ~ 50	90 ~ 60	110 ~ 70	1.5	2	3	4	6	8	11	15
30 ~ 25	35 ~ 30	40 ~ 35	50 ~ 40	60 ~ 50	70 ~ 60	2	3	4	6	8	12	18	20
25 ~ 20	30 ~ 25	35 ~ 30	40 ~ 35	50 ~ 40	60 ~ 50	3	5	6	9	12	18	25	30
20 ~ 15	25 ~ 20	30 ~ 25	35 ~ 30	40 ~ 35	50 ~ 40	5	8	10	15	20	25	30	30
15 ~ 10	20 ~ 15	25 ~ 20	30 ~ 25	35 ~ 30	40 ~ 35	12	18	18	25	30	30	30	30
10 ~	15 ~	20 ~	25 ~	30 ~	35 ~	25	30	30	30	30	30	30	30

Source: [1]

3. Results and discussion

Initial moisture content in all species woods investigated varied from the outside (near the cambium tissue) to the pith (center portion) of the stem (Table 5). The moisture gradient in wood is one of the most important physical parameters which can minimized internal stress and provide good dimensional stability (8).

Table 5. Types of defects and drying condition of each wood investigated.

Species	Initial moisture content (%)	T/R ratio	Type of defects			Initial Temp. (°C)	WBD (°C)	Final Temp. (°C)
			I	II	III			
Jaha	52 - 83	1.9	2	2 - 3	2	55	4	85
Kiacret	109 - 145	1.7-2.2 (1.8)	2 - 3	6 - 7	4 - 5	40	3	50
Kipasang	41 - 60	2.3-3.1 (2.6)	6 - 7	6 - 7	5 - 6	40	3	50
Red Meranti	53 - 58	1.8-2.1 (2.0)	2	2 - 3	2	55	4	85
White Meranti	52 - 80	2.5-3.0 (2.6)	4 - 5	5 - 6	3 - 5	40	3	65

Bira-Bira	99 - 112	2.6- 2.8 (2.7)	5	5 - 6	5	40	3	65
Bipa	57 - 71	1.9 – 2.3 (2.2)	3 - 4	2 -3	2 - 3	50	4	77

Remarks: I = end & surface check; II = deformation; III = honeycomb defect; 1 = very good; 2 = good; 3 = rather good; 4 = fair; 5 = rather poor; 6 = poor; 7 = very poor.

The initial moisture content in all wood species investigated varied from the outside to the pith of the stem, but it was still above fiber saturation point (Table 5). The high variability in initial moisture content at wood may be caused by its anatomical structure, reflected mainly in the variabilities in heartwood content and wood density [9]. The research result of drying five lesser-known woods species from Riau [10] showed that the variation in the initial moisture content of each wood species produces drying defect such as surface check, deformation on cross-section and honeycombing in various levels. At the same species, the wetter the wood is, the higher the difference between tangential and radial (T/R) shrinkage. T/R shrinkage values high, indicating its poor dimensional stability.

Early in the quick-drying test (at 100 °C temperature), when the moisture content is high, the fibers in the outer board dry first and begin to shrink. However, the inner has not yet begun to dry and shrink; consequently, the inner prevents the outer board from shrinking fully. Thus, the outer goes into tense and the inner into compressed causing a defect of the board, i.e. surface check, internal check/honeycombing, and spool-like deformation on the cross-section of the board. Surface check usually early in drying step and it occurs because drying stresses exceed the tensile strength of the wood perpendicular to the grain of the board, while honeycombing is internal crack caused by a tensile failure across the grain of the wood and usually occurs in the wood rays [1]. The failures can be the extended result from the surface and/or end checks. The research results of (11), showed the presence of honeycombing or internal check in wood could weaken its strength. So that it affects the quality and the value of wood. Deformation on the cross-section of the board can be caused by differences between radial and tangential shrinkage in the board when it dries, also aggravated by irregular grain and the presence of abnormal tree growth, such as juvenile and reaction wood [1]. Any deformation accompanied by the presence of internal checks (honeycomb defects) and wavy surface indicates the formation collapse where the wood cells are flattened due to excessive stress generated in the cells [3]. Severe deformation on the cross-section of the board or collapse must be avoided because it is a serious defect. The presence of these defects in dry wood is difficult to be processed further, except to be used as a specific product with artistic elements.

Based on the quick drying test results, it can be seen that surface check, deformation, and honeycomb defect was found in all species but with different degrees (Table 5). Jaha and Red Meranti wood have the best drying properties of the seven wood investigated,. The tendency for Jaha and Red Meranti wood to develop a surface check, deformation, and honeycomb defects during quick-drying test process were lower compared with the rest of the woods species. The woods species had the lowest shrinkage level, with a T/R ratio of ≤ 2. In general the smaller the difference between T/R-ratio in shrinkage, the smaller the degree of wood defects, like deformation and honeycomb [4,12,13]. Based on their drying properties in the quick-drying test, the drying condition for these wood species is at a dry-bulb temperatures range of 55 to 85 °C and initial wet-bulb depression of 4 °C.

Table 5 shows, kipasang woods had T/R shrinkage value 2.6 was categorized as difficult to dry. The wood tends to develop severe deformation and honeycomb during the drying process. Kipasang wood is difficult to dry because the density of the wood is high, which is an average of 0.89 g/cm^3 and there is a possibility of crystal in the parenchyma. If water is forced out of the wood using high temperature and low humidity at the beginning of the drying process, the wood will be damaged. Kiacret wood in the investigated had T/R shrinkage value 1.8, but the wood was difficult to dry. Although the T/R shrinkage ratio on Kiacret wood is low, it is difficult to dry. This is because of the possibility of tylosis in the wood vessels, thus inhibiting the process of removing water from the wood. For this reason, tylosis must be removed before the wood drying process. The process of removing tylosis from the wood vessels can be done using steam or boiling. Based on their drying properties in the quick-drying test, the drying condition for Kiacret and Kipasang are at a dry-bulb temperatures range of 40 to 50 °C and initial wet-bulb depression of 3 °C. Bipa wood, although the T/R shrinkage ratio slightly exceeds 2, but is tangential shrinkage is very small, only about 3%. This is causes the structure of the wood cell walls to be strong even when subjected to high temperature. Therefore, an investigation on basic properties in wood drying of Bipa wood is worth continuing in more detail, with respect to especially its relationship with the structure and or chemical composition of the wood. Based on their drying properties in the quick-drying test, the drying condition for Bipa wood is at a dry bulb temperatures range of 50 to 77 °C and initial wet-bulb depression of 4 °C.

Referring to the drying schedule as presented in Tables 3 and 4, as well as to the difference between wet and dry bulb temperature Table (1), the basic drying schedule for each wood species, which has 1-inch thick timber is presented in Table 6-9. However, modifying such drying schedules are still possible, depending on the drying-kiln type and thickness of the timber.

Table 6. Basic drying schedule for Jaha and Red Meranti wood.

Moisture content, %	Temperature, °C	Humidity, %
Green–30	55	81
30–25	60	73
25–20	65	64
20–15	70	47
15–10	80	30
≤10	85	24

Table 7. Basic drying schedule for Bipa wood.

Moisture content, %	Temperature, °C	Humidity, %
Green–30	50	80
0–25	55	72
25–20	60	62
20–15	65	45
15–10	80	30
≤10	80	22

Table 8. Basic drying schedule for White Meranti and Bira-Bira wood.

Moisture content, %	Temperature, °C	Humidity, %
Green–30	40	83
30–25	45	78
25–20	50	71
20–15	55	57
15–10	65	38
≤10	65	14

Table 9. Basic drying schedule for Kiacret and Kipasang wood.

Moisture content, %	Temperature, °C	Humidity, %
Green–30	40	83
30–25	42	77
25–20	42	77
20–15	45	69
15–10	50	54
≤10	50	29

4. Conclussion

Among the seven species investigated for their drying properties, Jaha and Red Meranti woods have the best drying properties and low tendency to develop defects, while Kiacret and Kipasang have the poor drying properties, and tend to develop severe deformation and honeycomb during their drying process.

Based on the drying condition of each wood, which obtained in the quick-drying test process showed that basic drying schedule of Jaha and Red Meranti wood required to achieve below 10% were at 55– 85 °C in temperature and 81–24% in relative humidity, Bipa wood were 50–65 °C and 83–17%, and for both White Meranti and Bira-Bira wood were 40-65 °C and 83–29%. Meanwhile, the corresponding temperature and relative humidity regarding the drying schedule for both Kiacret and Kipasang were 40–50 °C and 83–29% in relative humidity. However, it is still possible by modifying the drying schedule which considers the condition of drying kiln and dimensional sizes of board/lumber.

Acknowledgments

The authors gratefully acknowledge the support from the Forest Products Research and Development Center, Research Development and Innovation Agency, for financial support. We also thank to Abdurachman, ST. for his advice and suggestions.

References

[1] Simpson WT. *Drykiln Operator's Manual: Drying defects*. Simpson, editor. S. Department of Agriculture, Forest Prod. Laboratory. Agric. Handbook 188, Madison , Wisconsin. 1991. Pp 179-205.

[2] Bergman R. *Drying and Control of Moisture Content and Dimensional Changes*. In: Ross RJ, editor. Wood handbook: Wood as an Engineering Material. Centennial. Madison: U.S. Dept. of Agriculture, Forest Service, Forest Products Laboratory; 2010. p. 1–19.

[3] Yuniarti K, Ozarska B, Brodie G, Harris G and Waugh G. Collapse development of *Eucalyptus saligna* under different drying temperatures. *Journal of Tropical Forest Science*. 2015;27(4):462–71.

[4] Yuniarti K, Basri, E and Abdurachman. Honeycombing and deformation of six wood species and their relationship with several physical properties. *Jurnal Penelitian Hasil Hutan*. 2017;35(2):115–22.

[5] Effah B. The susceptibility of *Cola nitida* and *Funtumia elastica* to some drying defects. *International Journal of Science and Technology*. 2014;4(2):37–40.

[6] ASTM. *Standard Method of Testing Small Clear Specimen of Timber D143* (Modification). In: Annual Book of ASTM Standards. Volume 4.1. ASTM International, West Conshohocken.; 2012.

[7] Basri E. *The Quality of Waru Gunung (Hibiscus macrophyllus Roxb.) Wood at Three Age Group and its Densification for Furniture Material*. Gadjah Mada University; 2011.

[8] Yu L, Hao X, Hao, Cai L, Shi SQ, Jiang J and Lu J. An Investigation of Moisture Gradient in Wood during Drying Using X-Ray Radiation and Numeric Methods. *Forest Prod. Journal.* 2014;64(5/6):199–205.

[9] Panshin, AJ and de Zeeuw C. *Textbook of Wood Technology*. 4th ed. New York: McGraw-Hill Book Company.; 1980. 722 p.

[10] Yuniarti K and Basri E. High temperature drying properties and basic drying schedule of 5 lesser-known species from Riau. *Jurnal Penelitian Kehutanan Wallacea*. 2017;6(2):91–9.

[11] Phonetip K, Ozarska, B and Brodie GI. Comparing two internal check measurement methods for wood drying quality assessment. *European Journal of Wood and Wood Products*. 2017;75(1):139–42.

[12] Bowyer JL, Shmulsky R and Haygreen JG. *Forest Products & Wood Science: An Introduction*. 5th Edition. United Kingdom: Oxford, UK: Blackwell Publishing Ltd; 2007. 539 p.

[13] Basri E, Yuniarti K, Wahyudi I, Saefudin and Damayanti R. Effects of girdling on wood properties and drying characteristics of Acacia mangium. *Journal of Tropical Forest Science*. 2015;27(4):498–505.

Characteristics of rattan composite lumber made of *Calamus zollingeri*

I M Sulastiningsih*, **D R Trisatya and Sukadaryati**

Forest Products Research and Development Centre
Research, Development and Innovation Agency
Ministry of Environment and Forestry, Republic of Indonesia
Jl. Gunung Batu 5, Bogor, Indonesia.

*e-mail: ignasiasulastin@gmail.com

Abstract. To increase the efficiency of rattan utilization, the effort has been conducted by producing rattan composite lumber from unused big diameter rattan. The objective of the study was to determine the effects of layer compositions on the properties of rattan composite lumber (RCL). Rattan strips for RCL fabrication were prepared from *Calamus zollingeri*, bamboo strips were prepared from *Gigantochloa robusta and Gigantochloa pseudoarundinaceae*, while wood veneer was prepared from *Melia azedarach.* Laboratory scale of four kinds of RCL, i.e. rattan bare core (RBC), laminated rattan mayan bamboo lumber (MRM), laminated rattan andong bamboo lumber [ARA], and laminated rattan veneer lumber (VRV) with the dimension of 50 cm x 15 cm x thickness of RCL were fabricated using isocyanate adhesive. Results showed that the properties of RCL were significantly affected by layer compositions. Bamboo strips and wood veneer lamination on the surfaces of RBC improved the physical and mechanical properties of RCL. The improvement in thickness swelling of RCL varied from 35% (ARA)-75.7% (VRV), whereas the improvement in MOR, MOE and compression strength of RCL varied from 6.7% (VRV)-54.4% (ARA), from 40.9% (VRV)-276.8% (ARA) and from 12.8% (VRV)-54.1% (ARA), respectively. Laminated rattan-bamboo lumber and laminated rattan veneer lumber were suitable for furniture material.

1. Introduction

It was reported that in 2014 rattan products export from rattan handicrafts and furniture was USD 214.8 million [1]. Nevertheless, rattan export performance between 2012-2017 showed declining trend at around 1.7% per annum and in January-February 2015 it dropped to USD 27.7 million [1]. Even though raw [unwashed sulphurized] rattan export reached 904,905 culms in 2016, imported rattan was still desired to fulfill local rattan industries' needs [2].

Rattan canes are harvested when they are at maturation age, 6-8 years old and 12-15 years old for small and large diameter rattan species, respectively [3]. Rattan canes are harvested by cutting the stems using a long chopper and then pull down out the canopy [4]. It was cut about 1 m from the ground to give the chance to produce a new ramet [4]. Prior to manufacturing, harvested rattan are cleaned from the sheath, sanded and washed [3].

Small diameter rattan (< 30 mm) of less than five rattan species is commonly used in rattan processing manufactures. Large diameter rattan is less desirable in rattan processing manufacture compared to small diameter rattan species since the latter species is easier to manufacture. Thus, large diameter rattan species is less utilized and become a waste. Nevertheless, due to its lignocellulosic and other chemical content, rattan waste can be utilized as composite products [5]. Lignin content in rattan that was obtained from Klason lignin test was 21.2% [6] [7]. Lignin enhances cell wall rigidity and responsible for mechanical support [8]. However, rattan has less lignin content in comparison to hardwood (13-34%) and softwood (14-34%) [9].

Besides chemical properties, physical and mechanical properties play important roles in bearing load, thus they should be considered in utilizing rattan. Thickness swelling of rattan binderless board was above the standard of 12% that was 43.1% [6]. This number is however, superior to coconut husk and sugarcane bagasse binderless boards with values of 77.6% and 95.4%, respectively [6]. It was reported that *Calamus zolingerii* is one of the large diameter rattan species that has a durable bearing capacity for the furniture frame [10]. *C. zolingerii* is classified in to Class III (moderate) for its density (0.43-.0.52), class II (good) for its MOE and MOR [10].

One of the urgencies of the composite products development is to escalate the inferior properties of solid lignocellulosic material [11]. Laminated bamboo combined with wood veneer has demonstrated greater physical properties compared to solid wood [12]. Bamboo faced board satisfied the JIS A5908-2003 standard [13] [14]. Mayan and andong bamboo has been widely used as composite products [15] [16] [17] [18].

The density of laminated mayan and andong bamboo lumber were 0.6 g/cm^3 and 0.77 g/cm^3, respectively, while the density of sanded mayan and andong bamboo zephyr were 0.58 g/cm^3 and 0.73 g/cm^3, consecutively [15]. It was concluded that physical and mechanical properties of laminated bamboo lumber were influenced by bamboo species, with the exception to moisture content, compression strength parallel to grain and bonding strength [15].

Veneer can be manufactured from fast growing species, such as Mindi [*Melia azedarach*]. Mindi has been cultivated in Perum Perhutani Unit III West Java to address increasing market demand in the wood industry [19]. It is also planted in agroforestry system [20] and reforestation [19] [21]. Mindi has a smooth and shiny wood surface [22]. Besides its fancy look, the strength class of Mindi and its easiness in machining make it appropriate for furniture and panel production [19] [20] [22].

Mindi plywood adhered with urea-formaldehyde and phenol-formaldehyde had better static bending and shears strength properties than *Pinus taeda* plywood [23]. Thus, it was recommended for internal and external applications [23]. It was also reported that mindi plywood has remarkable physical and mechanical properties than kapuk [*Bombax ceiba*] plywood [20]. The density of mindi plywood was 541 kg/m^3 which was higher than its solid density of 430 kg/m^3 [20].

This research was conducted to determine the effect of layer compositions on the properties of rattan composite lumber (RCL). It studied the suitability of using rattan bare core, andong bamboo, mayan bamboo and mindi veneer as the cover layer of rattan strips.

2. Materials and methods

2.1. Materials

Batang rattan [*Calamus zollingeri* Beccari] strips were provided by PIRNAS [National Rattan Innovation Centre] in Palu, South Sulawesi, 200 cm long, 2 cm wide, and 2 cm thick. The bamboo materials used were andong bamboo (*Gigantochloa pseudoarundinacea* (Steudel) Widjaja) and mayan bamboo (*Gigantochloa robusta* Kurz) obtained from private gardens in Sukabumi, West Java, whereas wood material used was mindi (*Melia azedarach* L.) obtained from private garden in Bogor, West Java. Boron solution was used as preservative and water based polymer-isocyanate (WBPI) was used as adhesive in the production of rattan composite lumbers.

2.2. Preparation of rattan and bamboo strips materials

Rattan strips obtained from PIRNAS (200 cm long, 2 cm wide, and 2 cm thick) were cross cut into 50 cm in length. The strips were then immersed in 7% boron solution for two hours and after which they were air dried for one night and then followed by sun-dried to about 12% moisture content.

Bamboo strips were obtained by splitting bamboo bolts using bamboo splitter machine. Only the middle part of mayan bamboo and andong bamboo culms were used for this study. The bamboo culms were cross cut into 110 cm in length of the bamboo bolt and generally had two internodes. Six to eight bamboo strips of 2 cm in width were obtained from each bolt. Only straight bamboo strips were used for this study. The epidermis of bamboo strips was planed out to obtain the targeted thickness of 5 mm and the resulted bamboo strips were stacked for air drying at room temperature for one week. The bamboo strips were then immersed in 7% boron solution for four hours and after which they were air dried for one night and then followed by sun-dried to about 12% moisture content. The bamboo strips were then cross cut into 50 cm in length.

2.3. Preparation of wood veneer

Small diameter log (28 cm in diameter and 125 cm in length) of Mindi was converted into wood veneer with targeted thickness of 1.5 mm. The veneers were cut into certain dimension of 120 cm in length and 100 cm in width and were then air dried at room temperature for one week. The veneers were then further cross cut into 50 cm in length and 16 cm in width and after which they were sun dried to 12% moisture content.

2.4. Rattan bare core manufacture

Eight rattan strips were assembled side-by-side and edge-glued using water based polymer-isocyanate adhesive. The glue mix (main fluid 100 and cross linker 15) of 250 g/m^2 for a single glue line was then hand-spread on each thickness surface of rattan strips using a metal spatula. The assemblies were cold-pressed for one hour using a wooden clamp.

2.5. Bamboo sheet manufacture

Each bamboo sheet comprised of eight bamboo strips. The bamboo strips were assembled side-by-side and edge-glued using water based polymer-isocyanate adhesive. The glue mix [main fluid 100 and cross linker 15] of 250 g/m^2 for a single glue line was then hand-spread on each thickness surface of bamboo strips using a metal spatula. The assemblies were cold-pressed for one hour using a wooden clamp.

2.6. Rattan composite lumbers (RCLs) manufacture

RCLs were produced by assembling rattan bare core (RBC) covered by bamboo sheet or wood veneer. Four types of RCL were produced with four different layer compositions, i.e., rattan bare core (RBC), rattan bare core in which the top and bottom layers covered with mayan bamboo sheets and produced laminated rattan-mayan bamboo-lumber (MRM), rattan bare core in which the top and bottom layers covered with andong bamboo sheets and produced laminated rattan-andong bamboo-lumber (ARA) and rattan bare core in which the top and bottom layers covered with wood veneers and produced laminated rattan-veneer-lumber (VRV). The laboratory scale of 3-layer rattan composite lumbers with the dimension of 50 cm x 15 cm x thickness of RCL were manufactured using water based polymer-isocyanate (WBPI) adhesive. The assemblies were cold-pressed for one hour and the glue spread applied was 250 g/m^2. Three replications for each treatment of RCL were prepared. The RCLs produced were conditioned for two weeks before testing.

2.7. Testing and data analysis

The RCLs were trimmed, cut into desired specimen dimensions and measured for density, moisture content, water absorption, thickness swelling, delamination, static bending (modulus of rupture (MOR) and modulus of elasticity (MOE)), and compression strength. The tests of density, moisture content, thickness swelling, and water absorption were performed using the American Standard ASTM D 1037-93, whereas the tests of compression strength were performed using ASTM D 143-94. The bonding quality (immersion delamination test) and static bending (MOR and MOE) tests were performed in accordance with the Japanese Standard for Glued Laminated Timber (JPIC, 2003). A completely randomized design was used in the experiment with the layer compositions as the treatment factor. Three replications were prepared for each treatment.

3. Results and discussion

3.1. Physical properties of rattan composite lumbers

The effects of layer compositions on the physical properties of rattan composite lumber are discussed below. The mean values of physical properties of rattan composite lumbers are presented in Table 1. The data on physical properties of rattan composite lumbers were subjected to analysis of variance (ANOVA) and the results are presented in Table 2.

Table 1. Physical properties of rattan composite lumber.

Properties	Layer compositions of rattan composite lumber			
	RBC	MRM	ARA	VRV
Moisture content, %	11.96 (0.29)	12.16 (0.11)	12.14 (0.19)	12.24 (0.32)
Density, g/cm^3	0.305 (0.011)	0.483 (0.004)	0.435 (0.003)	0.344 (0.013)
Water absorption, %	125.13 (1.78)	61.87(0.49)	78.46 (1.29)	58.26 (4.72)
Thickness swelling, %	4.86 (0.54)	2.57 (0.39)	3.16 (0.45)	1.18 (0.13)
Delamination, %	0	0	0	0

Remarks: RBC= rattan bare core; MRM= rattan bare core covered with mayan bamboo sheets; ARA= rattan bare core covered with andong bamboo sheets; VRV= rattan bare core covered with mindi wood veneers; Each value was the average of three speciments; Values in parentheses are standard deviations.

Table 2. Summary of ANOVA on physical properties of rattan composite lumber.

Properties	F –Value	P-Value
Moisture content	0.72^{ns}	0.566
Density	154.74**	0.000
Water absorption	414.83**	0.000
Thickness swelling	42.09**	0.000

Remarks: ^{ns} not significant; ** highly significant

The moisture content of rattan composite lumbers varied from 11.96% to 12.16% with an average of 12.13%. These values meet the Japanese Standard requirement for glued laminated timber since the values are below 15%. The result of ANOVA in Table 2 showed that the moisture content of rattan composite lumber was not affected by layer compositions. The moisture content of dried rattan was around 8 to 20% that indicated the surrounding humidity [3].

The density of rattan composite lumbers in this study varied from 0.305 g/cm^3 to 0.483 g/cm^3 with an average of 0.392 g/cm^3. The results of ANOVA in Table 2 showed that the density of rattan composite lumber was significantly affected by layer compositions. Rattan bare core covered with mindi veneer had the highest density, while rattan with the absence of bamboo or wood veneer layers had the lowest density [Table 2]. It was reported that the density of mindi plywood was 0.541 g/cm^3 [23], andong and mayan bamboo were 0.88 g/cm^3 and 0.83 g/cm^3, respectively [24]. The density of sanded bamboo strips was 0.73 g/cm^3 and 0.58 g/cm^3, respectively for andong and mayan bamboo [15]. Even though mindi had the smallest density value among mayan and andong bamboo, when it was composed with rattan bare core it constructed the highest density. This might be due to the higher extractive content of mindi in comparison to andong and mayan bamboo. Extractive content of wood is one of the influencing factors of wood density [11] [25]. The previous study revealed that the density of *Tristaniopsis whiteana* and *T.beccarii* significantly decreased as they were extracted [25]. The cellulose content in rattan, however, is higher compared to hardwood and softwood [3]. Also, wood density is inversely related to wood porosity or the void volume of wood [11]. Results in this study showed that relatively low density of rattan, however, can be improved by producing composite products using wood veneers or laminated bamboo. This echoes the earlier findings of [23] that suggested to apply gluing technologies for the production of re-formed wood products from low density wood.

Water absorption of rattan composite lumbers varied from 58.26% to 125.13% with an average of 80.93%. The results of ANOVA in Table 2 showed that water absorption values were significantly affected by the layer compositions. It was observed that water absorption of rattan composite lumber decreased as laminating rattan bare core either with bamboo sheet or wood veneer. The highest water absorption was rattan bare core [125.13%] while the lowest water absorption was rattan bare core covered with mindi veneer [58.26%]. Another earlier study revealed that mindi plywood [63.67%] has better water absorption value than kapok plywood (107.32%) [20]. Besides its porosity, water absorption is also influenced by board density [26] [27]. The previous study on the properties of strandboard bamboo showed that low density board experienced more water absorption than that of high density board [27]. This might due to the high compaction ratios of high density boards that enabled to absorb more water than low density board [28]

Thickness swelling is one indicator of dimensional stability. It performs as an important property of rattan composite lumber. The thickness swelling of rattan composite lumber varied from 1.18% to 4.86% with an average of 2.94%. The results of ANOVA in Table 2 showed that thickness swelling of rattan composite lumber was significantly affected by layer compositions. It was observed that the values of thickness swelling decreased with covering rattan bare core with bamboo sheet or wood veneer. Mindi veneer showed the superior physical property of thickness swelling than other layers, i.e. 1.18%. Compared to kapok plywood, mindi plywood was also had more stable dimension, where thickness swelling of kapok plywood was 3.16% and mindi plywood was 2.51% [20]. It has been shown that the thickness swelling values of laminated andong and mayan were 3.31% dan 4.57%, consecutively [15]. Moreover, it was concluded that laminated bamboo strips were more stable than laminated bamboo zephyr [15]. The board density influenced thickness swelling of bamboo strandboard for short-term soaking [27]. Furthermore, it was suggested that thickness swelling was also be affected by resin type, the internal structure of the fiber, temperature [27] [6], the hygroscopicity characteristics of hemicellulose [6] and the presence of extractives [29].

3.2 Mechanical properties of rattan composite lumbers

The mean values of the mechanical properties of rattan composite lumbers are presented in Table 3. The data on mechanical properties of rattan composite lumbers were subjected to analysis of variance (ANOVA) and the results are presented in Table 4.

The MOR of rattan composite lumber (Table 3) varied from 312.6 kgf/cm^2 (30.6 MPa) to 482.8 kgf/cm^2 (47.3 MPa) with an average of 384.7 kgf/cm^2 (37.7 MPa). The results of ANOVA in Table 4 showed that the MOR of rattan composite lumber was significantly affected by layer compositions. Rattan bare core covered with andong bamboo shown the highest MOR [482.8 kgf/cm^2] compared to mayan bamboo (408.4 kgf/cm^2) and mindi veneer (335 kgf/cm^2). Furthermore, the absence of external layers in rattan composite lumber gave the inferior properties of MOR (312.6 kgf/cm^2).

Plywood panels entirely made of mindi presented superior MOR compared to kapok plywood, ie. 58.33 N/mm^2 or 594.80 kgf/cm^2 and 32.52 N/mm^2 or 331.61 kgf/cm^2, respectively [20]. Comparative findings from this earlier study indicate that composite lumber made of rattan bare core and mindi veneer had lower MOR than mindi plywood. It was reported that *Calamus zollingeri* stem contains around 41.1% of cellulose, 21.2% of lignin, 1.4% of silica and 20.6% of starch [30]. MOR value is highly influenced by wood/lignocellulosic density and cellulose content [6].

The previous study [31] revealed that two layers of andong bamboo laminate had 1,339 kgf/cm^2 of MOR, while the MOR for *Calamus zollingerii* was 580 kgf/cm^2 [30]. The two layers bamboo laminate had greater MOR value than rattan composite lumber with external layers of andong [482.8 kgf/cm^2] and mayan bamboo (408.4 kgf/cm^2) that were assembled in this study. Another study has revealed the MOR values for three layers bamboo laminate were 1,236.7 kgf/cm^2 and 1,107.24 kgf/cm^2 for andong and mayan, consecutively [15]. This might be due to the higher density of bamboo compared to rattan. The density of andong and mayan bamboo was 0.88 g/cm^3 and 0.83 g/cm^3, respectively [24], while the specific gravity of rattan was 0.49 [30]. The difference of vascular bundle type between andong and mayan might be contributed to MOR value, as investigated in the comparative study of three species of bamboo [32].

MOE values of rattan composite lumber varied from 20,609 kgf/cm^2 or 2,020.5 MPa to 77,651 kgf/cm^2 or 7,612.8 MPa with an average of 49,443 kgf/cm^2 or 4,847.4 MPa. ANOVA in Table 4 showed that the MOE of rattan composite lumber was significantly affected by the layer compositions. The RBC had the lowest MOE value (20,609 kg/cm^2) compared to others. The MOE of *C. zollingerii* cane in the other study, however, was 29,442 kg/cm^2 [30]. The slightly different value of MOE might be due to the removal of silica epidermis from *C. zollingerii* cane in the preparation of rattan strips.

Table 3. Mechanical properties of rattan composite lumbers.

Properties	Layer compositions of rattan composite lumber			
	RBC	MRM	ARA	VRV
Modulus of rupture (MOR), kgf/cm^2	312.6 (23.9)	408.4 (54.2)	482.8 (43.6)	335.1 (11.91)
Modulus of elasticity (MOE), kgf/cm^2	20,609 (921)	70,482 (2,438)	77,651 (4,777)	29,033 (488)
Compression strength, kgf/cm^2	173.62 (3.35)	261.57 (9.89)	267.51 (9.72)	195.92 (14.05)

Remarks: RBC= rattan bare core; MRM= rattan bare core covered with mayan bamboo sheets; ARA= rattan bare core covered with andong bamboo sheets; VRV= rattan bare core covered with mindi wood veneers; Each value was the average of three speciments; Values in parentheses are standard deviations.

Table 4. Summarized ANOVA of mechanical properties of rattan composite lumbers

Properties	F -Value	P-Value
Modulus of rupture [MOR]	12.85**	0.002
Modulus of elasticity [MOE]	333.18**	0.000
Compression strength	66.17**	0.000

Remarks: ** highly significant

The use of mindi veneer as external layers of rattan bare core improved the elasticity of the lumber with the value of MOE 29,033 kgf/cm^2. However, it was lower than that of mindi plywood that was produced in the preceding research (3,950.01 N/mm^2 or 40,278.89 kgf/cm^2) [20]. Andong rattan composite lumber presented superior MOE value compared to mayan and mindi composite lumber (Table 3). Two layers andong lamina gave the highest MOE value (222.834 kgf/cm^2) when it was assembled in outer-outer direction [31]. Additionally, andong bamboo internodes gave a higher value of MOE than bamboo nodes due to the higher proportion of vascular bundle of the former part [31]. MOE of three layers andong lamina was higher (361.154 kgf/cm^2) than mayan lamina (143.127 kgf/cm^2) [15]. It confirms that the additional number of layers of bamboo improved elasticity of the composite.

The compression strength of rattan composite lumbers varied from 173.62 kgf/cm^2 or 17.03 MPa to 267.51 kgf/cm^2 or 26.23 MPa with an average of 195.92 kgf/cm^2 or 19.21 Mpa. The results of ANOVA (Table 4) showed that the compression strength of rattan composite lumbers was significantly affected by layer compositions. Rattan bare core with andong bamboo external layers had slightly greater compression strength value [267.51 kgf/cm^2] compared to mayan (261.57 kgf/cm^2). However, the compression strength of laminated mayan bamboo board with three and five layers were 571.98 kgf/cm^2 and 503.17 kgf/cm [16]. Thus, compression strength was not affected by the number of layers. The compression strength parallels to grain of laminated mayan and andong bamboo were 594.65 kgf/cm^2 and 700.88 kgf/cm^2, respectively.

Table 5. Post hoc test results on physical and mechanical properties of RCLs

Properties	Layer compositions of rattan composite lumber			
	RBC	MRM	ARA	VRV
Moisture content, %	11.96 a	12.16 a	12.14 a	12.24 a
Density, g/cm^3	0.305 a	0.483 b	0.435 c	0.344 d
Water absorption, %	125.13 a	61.87 b	78.46 c	58.26 c
Thickness swelling, %	4.86 a	2.57 b	3.16 b	1.18 c
Modulus of rupture, kgf/cm^2	312.6 b	408.4 ab	482.8 a	335.1 b
Modulus of elasticity, kgf/cm^2	20,609 d	70,482 b	77,651 a	29,033 c
Compression strength, kgf/cm^2	173.62 b	261.57 a	267.51 a	195.92 b

Remarks: RBC= rattan bare core; MRM= rattan bare core covered with mayan bamboo sheets; ARA= rattan bare core covered with andong bamboo sheets; VRV= rattan bare core covered with mindi wood veneers; Values followed by the same letter within the same row are not significantly different.

4. Conclusion

The study concluded that mayan bamboo (*Gigantochloa robusta* Kurz) strips, andong bamboo (*Gigantochloa pseudoarundinacea* (Steudel) Widjaja) strips and mindi (*Melia azedarach* L.) wood veneer could be used as surface layer material for rattan composite lumber production. Layer compositions showed a significant effect on the physical and mechanical properties of rattan composite lumber. Covering rattan bare core with bamboo strips or wood veneer improved dimensional stability and mechanical properties of rattan bare core.

Acknowledgment

The authors would like to thank Jasni for her continuous encouragement of undertaking this study.

References

1. Kementerian Perdagangan. Lampu kuning industri rotan indonesia. Jakarta; 2017.
2. Badan Pusat Statistik. Statistik Produksi Kehutanan 2016. Jakarta: Badan Pusat Statistik; 2017.
3. Rachman O, Jasni. Rotan: Sumberdaya, Sifat dan Pengolahannya. Bogor: Pusat Penelitian dan Pengembangan Keteknikan Kehutanan dan Pengolahan Hasil Hutan; 2013.
4. Hess B. Simulation of rattan harvests in Indonesia: Different harvesting pressures and the resulting patterns. Southern Swedish Forest Research Centre; 2013.
5. Ahmad Z, Tajuddin M, Salim NFF, Halim Z. Effect of alkaline treatment on properties of rattan waste and fabricated binderless particleboard. IIUM Eng J. 2018;19(1):185–96.
6. Ahmad Z, Tajuddin M, Wan Mohd Yusof W-SM. Mechanical, physical and thermal properties of rattan fibre based binderless board. J Trop For Sci. 2017;29[4]:485–92.
7. Maisarah T, Zuraida A, Marjan A. The effect of black liquor on rattan waste binderless board [BPB] produced via Digestion-Hot Pressing [DHP] process. ARPN J Eng Appl Sci. 2015;10(21):9753–8.
8. Liu Q, Luo L, Zheng L. Lignins: Biosynthesis and biological functions in plants. Int J Mol Sci. 2018;19(2).
9. Tsoumis G. Science and technology of wood. Structure, properties, utilization. Sci Technol wood Struct Prop Util. 1991;xiii + 494.
10. Abdurachman, Jasni. Penggolongan performans 25 jenis rotan Indonesia berdasarkan kerapatan, kekakuan dan kekuatan. J Penelit Has Hutan. 2016;33(4):273–82.
11. Shmulsky R, Jones PD. Forest Products and Wood Science An Introduction: Sixth Edition. Forest Products and Wood Science An Introduction: Sixth Edition. 2011.

12. Sulastiningsih IM, Santoso A, Krisdianto. Karakteristik balok bambu lamina susun tegak dari bilah bambu andong [Gigantochloa pseudoarundinacea [Steud.] Widjaja]. J Penelit Has Hutan [Internet]. 2016;34[3]:167–77. Available from: http://ejournal.forda-mof.org/ejournal-litbang/index.php/JPHH/article/view/2285
13. Khaeruddin CA. Pengaruh jenis lapisan luar terhadap kualias papan komposit. Institut Pertanian Bogor; 2008.
14. Setyawati D, Hadi YS, Massijaya MY, Nugroho N. Utilization of Coir and Recycled Polypropylene for Bamboo Matting Layer Composite. Mol Cryst Liq Cystals. 2008;484(1):389–400.
15. Sulastiningsih IM. Beberapa sifat bambu lamina yang terbuat dari tiga jenis bambu. J Penelit Has Hutan. 2008;26(3):277–87.
16. Sulastiningsih IM, Nurwati. Physical and mechanical properties of laminated bamboo board. J Trop For Sci. 2009;21(3):246–51.
17. Sulastiningsih IM, Santoso A. Pengaruh jenis bambu, waktu kempa dan perlakuan pendahuluan bilah bambu terhadap sifat papan bambu lamina. J Penelit Has Hutan. 2012;30[3]:199–207.
18. Abdurachman, Hadjib N, Jasni, Balfas J. Sifat Balok Komposit Kombinasi Bambu Dan Kayu. J Penelit Has Hutan. 2015;33(2):115–24.
19. Panggabean M. Struktur anatomi kayu mindi [Melia azedarach]. Universitas Sumatera Utara; 2008.
20. Rahman M, Asaduzzaman M, Rahman M, Das A, Biswas S. Physical and mechanical properties of ghora neem (*Melia azedarach*) plywood. Bangladesh J Sci Ind Res. 2014;49(1):47–52.
21. Idjuddin AA, Erfandi D, Soelaeman Y, Suwanda MH, Suganda H. Rehabilitasi dan konservasi tanah pasca erupsi Gunung Merapi. :123–36.
22. Martawijaya A, Kartasujana I, Mandang Y., Prawira SA, Kadir K. Atlas Kayu Indonesia Jilid II. Bogor: Pusat Penelitian dan Pengembangan Hasil Hutan; 2005. 167 p.
23. Iwakiri S, Trianoski R, De Castro VG, Parchen CFA, De Araujo RD. Quality of plywoods obtained from melia azedarach combined with pinus taeda, produced with urea-formaldehyde and phenol-formaldehyde resins. Floresta. 2014;44(1):93–100.
24. Barly, Ismanto A, Martono D, Abdurachman, Andianto. Sifat fisis dan stabilisasi dimensi beberapa jenis bambu komersial. J Penelit Has Hutan. 2012;30(3):163–70.
25. Abu Zaharin F binti. Effect of extractives on wood density and natural durability of Tristaniopsis whitenan and T. beccarii [myrtaceae]. 2013;
26. Sulastiningsih IM, Tambunan B, Yulia Y, Santoso A. Some properties of Bamboo-Cement Board. In: Proceedings of the 1st International Wood Science Seminar. 1996. p. 83–91.
27. Sumardi I, Suzuki S. Dimensional stability and mechanical properties of strandboard made from bamboo. BioResources. 2014;9(1):1159–67.
28. Kelly MW. Critical literature review of relationships between processing parameters and physical properties of particleboard [Internet]. Vol. 010, Gen. Tech. Rep. FPL-10. Madison, WI: U.S. Department of Agriculture, Forest Service, Forest Products Laboratory. 65 p. Madison; 1977 [cited 2018 Feb 27]. Available from: https://www.fs.usda.gov/treesearch/pubs/9820
29. Rowell R. Moisture Properties. Handb Wood Chem Wood Compos Second Ed [Internet]. 2012;75–98. Available from: http://www.crcnetbase.com/doi/10.1201/b12487-7
30. Jasni, Damayanti R, Kalima T. Atlas Rotan Jilid 1. Edisi 2. Rachman O, Mogea JP, editors. Bogor; 2012.
31. Augistyra DD. Distribusi ikatan pembuluh, sifat fisis mekanis bilah bambu dan bambu laminasi dua lapis. Institut Pertanian Bogor; 2012.
32. Nuriyatin N, Surjokusumo S. Korelasi antara pola ikatan pembuluh dengan sifat fisis dan mekanis tiga jenis bambu. J Ilmu dan Teknol Kayu Trop. 2010;8(2):133–44.

Lignocellulosic materials modification and engineering in relation to viscoelastic perspectives

W Dwianto[1*], A Bahanawan[1], T Darmawan[1], F Akbar[1], S Sufiandi[1], D S Adi[1], and R Damayanti[2]

[1] Research Center for Biomaterials, Indonesian Institutes of Sciences,
Jl. Raya Bogor KM 46, Cibinong, Bogor 16911, West Java, Indonesia
[2] Forest Products Research and Development Center,
Jl. Gunung Batu No. 5, Bogor 1660, West Java, Indonesia

[*]email: wahyudwianto@yahoo.com

Abstract. By 2010, only 52% or 94 million ha of the total land area in Indonesia was forested. Therefore, we should develop forest plantation with fast growing species, or better efficiency and quality improvement for the utilization of forest products. The conventional method of fabricating curved wooden products is to cut wooden beams into a curved wood, and to connect them for the expected curve shape. However, the disadvantages of this conventional woodworking process are wasting of wood raw materials, decreasing the wood strength due to fiber cutting, and reducing the beauty of wood fiber direction. These problems can be overcomed by bending solid wood, as well as curving laminated veneer lumber (LVL). The bending method of solid wood requires special technique and equipment compared to curving method of LVL. The development of commercial products is required to understand basic knowledge of wood bending. This paper deals with lignocellulosic materials modification and engineering in relation to viscoelastic perspectives; describing viscoelastic property of wood and other lignocellulosic materials, and development of furnitures and bicycle frames from LVL curves. It is expected that more modified and engineered products can be made from lignocellulosic materials based on the knowledge of their viscoelastic properties.

1. Introduction

The Indonesian archipelago of about 17,000 islands is a home to some of the most biodiverse forests in the world. In 1900, the total forest represented 84% of the total land area [1]. The forest cover by 1950 is estimated to 145 million ha of primary forest and another 14 million ha of secondary and tidal forest [2]. In the early 1970s, Indonesia used this valuable resource to its economic benefit with the development of the country's wood-processing industries. From the late 1980s to 2000, production capacity has increased nearly 700% in the pulp and paper industries, making Indonesia the world's ninth largest pulp producer and eleventh largest paper producer [1]. Between 1990 and 2000, 20% of the forest area in Indonesia had been lost (24 million ha) and by 2010, only 52% or 94 million ha of the total land area was forested [3]. Even despite a moratorium on new logging contracts imposed in 2010, the rate of deforestation continued to increase to an estimated 840,000 ha in 2012, surpassing deforestation in Brazil [4].

IOP Conf. Series: Earth and Environmental Science **359** (2019) 012013 doi:10.1088/1755-1315/359/1/012013

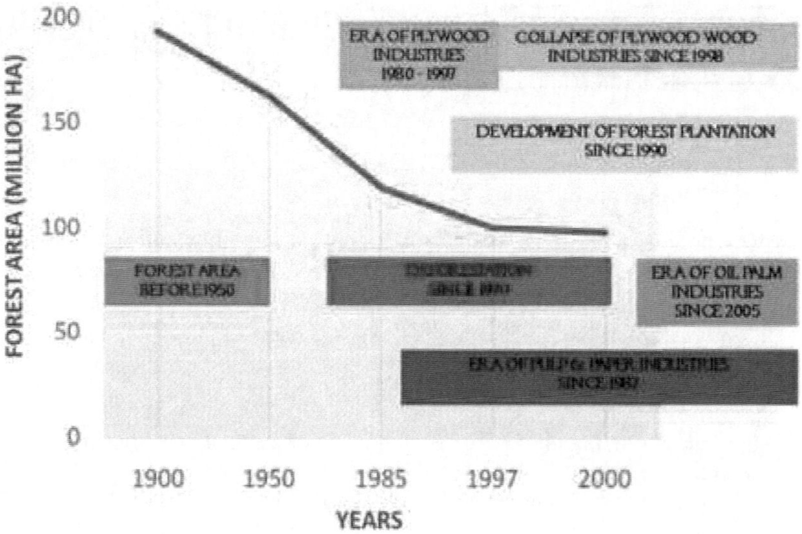

Figure 1. Decreasing of forest area in Indonesia.

After decreasing of natural forest area due to deforestation, following by collapse of plywood industries, Indonesia is now facing the era of forest plantation with fast growing species which mostly for pulp and paper industries, and oil palm plantation (Figure 1). Therefore, the more efficient and improved quality of forest products utilization is crucial.

2. Efficiency and quality improvement for the utilization of forest products

2.1. Conventional method of furniture manufacturing

Curved wooden products are generally applied to furniture products, as well as to residential building components (door frames/windows), musical and sport instruments, toys, and other necessities. The conventional method of obtaining curved wooden products is shaped by cutting the wooden beams into a curved wood, and connecting them to obtain the expected curve shape (Figure 2). Based on the process, this conventional method is easy to work, because it only uses simple wooden techniques and tools. However, the deficiency and disadvantages of this conventional woodworking process are wasting of wood raw materials, decreasing the wood strength due to fiber cutting, and reducing the beauty of wood fiber direction.

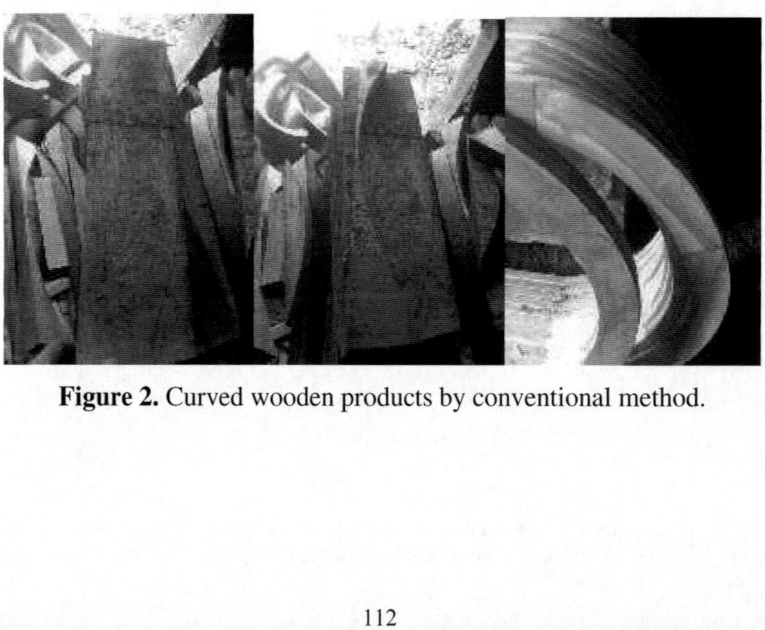

Figure 2. Curved wooden products by conventional method.

2.2. Research and development on wood bending and LVL curves

These problems can be overcomed by solid wood bending (Figure 3), as well as Laminated Veneer Lumber (LVL) curving (Figure 4) methods. Bending method of solid wood requires special technique and equipment compared to curving method of LVL, so that if it is to be developed commercially, it needs basic knowledge of wood bending.

Research Center for Biomaterials, Indonesian Institute of Sciences has been conducting research and development on wood bending technology since 2004 [5–8], and continuing with LVL [9–10]. The wood bending mechanism is almost the same as radial compression of wood [11], however the inner part of curve is compressed in longitudinal direction. Therefore, it can be done through its viscoelastic processes [12–14], i.e. (I) softening, it can be achieved by soaking the wood until water saturated and proceed with heating treatment; (II) forming, it may be possible when the wood is still soft when compared to dry wood; (III) setting, drying set is a drying process when wood is in a deformed condition; (IV) fixation, it is a particular effort/treatment for the wood that has been deformed does not return to its original shape or permanent.

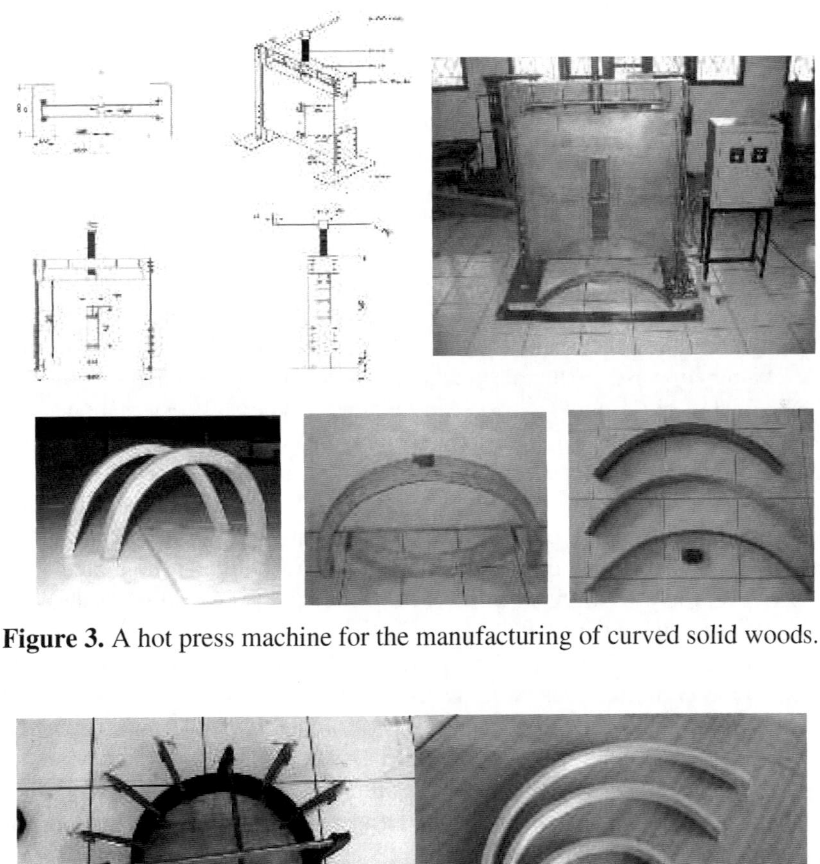

Figure 3. A hot press machine for the manufacturing of curved solid woods.

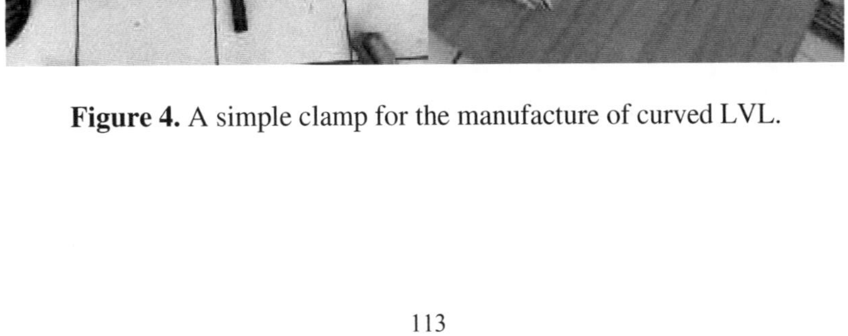

Figure 4. A simple clamp for the manufacture of curved LVL.

3. Viscoelastic property of wood and other lignocellulosic materials

3.1. Theoretical definition of viscoelastic materials

To bend wood or other lignocellulosic materials, their viscoelastic property should be known. Viscoelasticity is a property of materials that exhibit both viscous and elastic characteristics when they are undergoing deformation. Viscous materials resist shear flow and strain linearly with time when a stress is applied. Elastic materials strain when stretched, and immediately return to their original state once the stress is removed. Viscoelastic materials have elements of both of these properties, and thus exhibit time-dependent strain. Whereas elasticity is usually the result of bond stretching along crystallographic planes in an ordered solid, viscosity is the result of the diffusion of atoms or molecules inside an amorphous material [15].

Depending on the change of strain rate versus stress inside a material, the viscosity can be categorized as having a linear, non-linear, or plastic response. When a material exhibits a linear response, it is categorized as a Newtonian fluid. In this case, the stress is linearly proportional to the strain rate. If the material exhibits a non-linear response to the strain rate, it is categorized as Non-Newtonian fluid. In addition, when the stress is independent of this strain rate, the material exhibits plastic deformation [15]. Many viscoelastic materials exhibit rubber like behavior that is explained by the thermodynamic theory of polymer elasticity. Some examples of viscoelastic materials include amorphous polymers, semicrystalline polymers, biopolymers, and metals at very high temperatures. Cracking occurs when the strain is applied quickly and outside of the elastic limit.

Unlike purely elastic substances, a viscoelastic substance has an elastic component and a viscous component. The viscosity of a viscoelastic substance gives the substance a strain rate dependence on time. Purely elastic materials do not dissipate energy (heat) when a load is applied, then removed. However, a viscoelastic substance loses energy when a load is applied, then removed. Hysteresis is observed in the stress–strain curve, with the area of the loop being equal to the energy lost during the loading cycle. Since viscosity is the resistance to thermally activated plastic deformation, a viscous material will lose energy through a loading cycle. Plastic deformation results in lost energy, which is uncharacteristic of a purely elastic material's reaction to a loading cycle [15].

Specifically, viscoelasticity is a molecular rearrangement. When a stress is applied to a viscoelastic material such as a polymer, parts of the long polymer chain change positions. This movement or rearrangement is called creep under a contant stress. Polymers remain a solid material even when these parts of their chains are rearranging in order to accompany the stress, and as this occurs, it creates a back stress in the material. When the back stress hasthe same magnitude as the applied stress, the material no longer creeps. When the original stress is taken away, the accumulated back stresses will cause the polymer to return to its original form. The material creeps, which gives the prefix visco-, and the material fully recovers, which gives the suffix–elasticity [16].

The secondary bonds of a polymer constantly break and reform due to thermal motion. Application of a stress favors some conformations over others, so the molecules of the polymer will gradually "flow" into the favored conformations over time [17]. Because thermal motion is one factor contributing to the deformation of polymers, viscoelastic properties change with increasing or decreasing temperature. In most cases, the creep modulus, defined as the ratio of applied stress to the time-dependent strain, decreases with increasing temperature. Generally speaking, an increase in temperature correlates to a logarithmic decrease in the time required to impart equal strain under a constant stress. In other words, it takes less work to stretch a viscoelastic material an equal distance at a higher temperature than it does at a lower temperature. Both wood and cellulose materials are involved in such materials and the viscoelasticity which causes a variety of mechanical properties is closely related to thermal motions of a polymer chain of the material.

3.2. Bending strength of wood in softening condition

Viscoelastic property of wood can be simply measured by static bending test. Mechanical property changes due to the moisture content (MC) and/or temperature changes have been examined for 15 Indonesian wood species [18]. The densities ranged from 0.27 g/cm^3 for randu alas to 0.81 g/cm^3 for lamtoro wood.

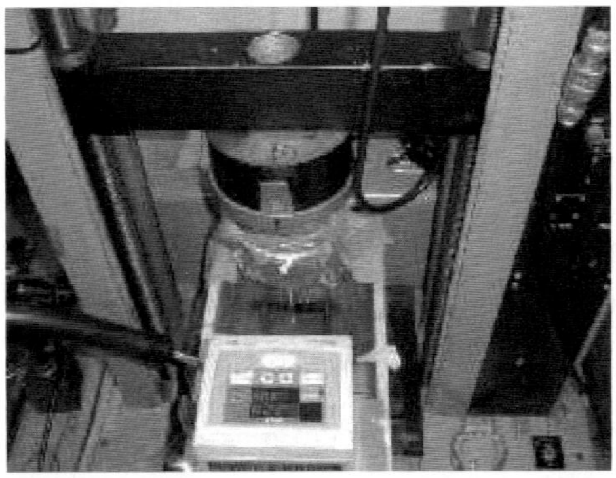

Figure 5. Static bending test in a water bath.

The test was conducted at 20°C and 65% RH using a conventional testing instrument with a constant cross-head speed (5 mm/min), loading at a three-point bending system with a span of 80 mm. In the water-saturated state, the load was applied to the specimen standing in a water bath controlled at 20°C and 80°C (Figure 5).

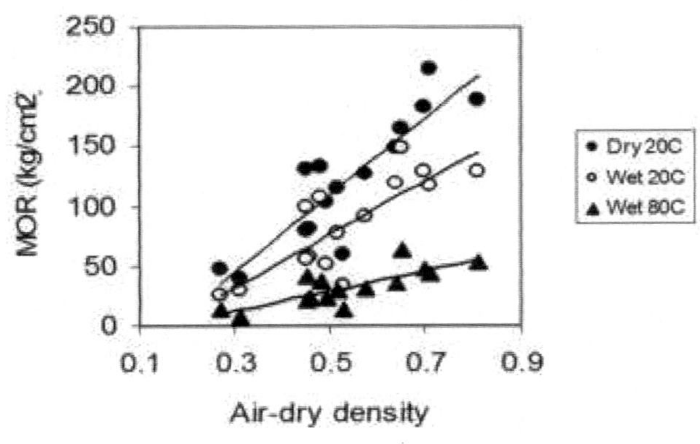

Figure 6. MOR of woods in softening conditions.

Figure 6 shows modulus of rupture (MOR) plotted against air-dry density. The MOR increased linearly with air-dry density, regardless of wood species. The wet 20°C/air-dry ratios for the slopes in regression lines was 0.67, while wet 80°C/air-dry ratios was 0.27 [18]. Iida [19] has reported that the wet 20°C/air-dry ratios was 0.62, and wet 80°C/air-dry ratios was 0.29 for Japanese wood species. The influences of MC change on MOR were somewhat smaller for the 15 Indonesian wood species than for the Japanese wood species. On the other hand, the influences of combined MC and temperature changes to MOR did not differ between Indonesian and Japanese wood species.

3.3. Relaxation modulus of wood in softening conditions

Figures 7 shows the relative relaxation modulus values of kecapi wood at air-dry 20°C, wet 20°C and wet 80°C. Relative relaxation modulus (ratio of relaxation modulus after treatment devided by initial relaxation modulus) decreased with an increase in the MC and temperature. The value of relative relaxation modulus in wet 20°C/air-dry ratios was decreased to around 0.8. However, it was affected still more by both MC and temperature, the wet 80°C/air-dry ratios of relaxation modulus decreased to around 0.6. It was concluded that the hygro-thermal properties of lignin significantly affect the changes of elastic and strength properties of wood that occur due to hygro-thermal effects.

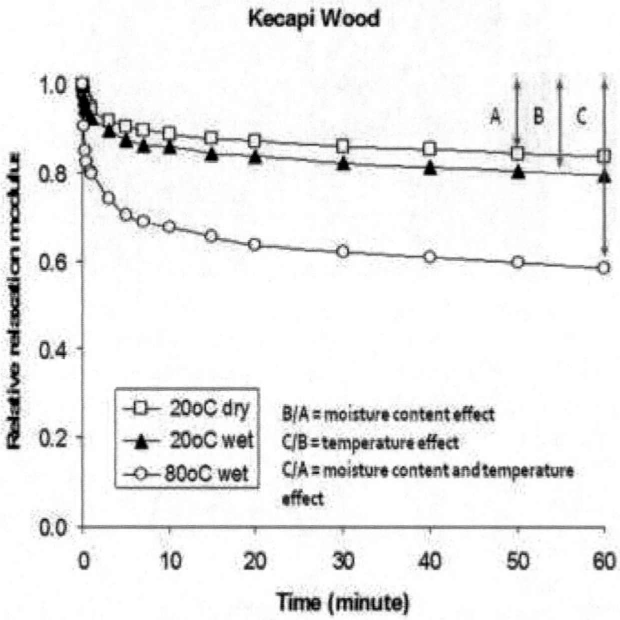

Figure 7. Relative relaxation modulus of kecapi wood in softening conditions.

3.4. Viscoelastic perspectives of other lignocellulosic materials

Furthermore, the softening behaviour and viscoelastic perspectives of other lignocellulosic materials, such as bamboo, rattan, and oil palm trunk was investigated. The specimens were tested at static bending in dry and wet conditions.

Figure 8 shows the decreasing of MOR of andong bamboo, sengon wood, air rattan and oil palm trunk from dry to wet condition. MOR of bamboo decreased drastically (y = –675.5x + 1984), followed by wood (y = –308.5x + 875.1), rattan (y = –195.9x + 608.2) and oil palm trunk (y = –65.19x + 386.7). It was suggested that besides the softening behaviour of their chemical components, bend-ability of lignocellulosic materials was depended on their anatomical structures. From the observations, the changes of anatomical structures occurred mostly in the inner parts of bamboo cell walls. This was due to the compression of the inner parts of bamboo in both longitudinal and radial directions, while in the middle parts and near the bark was remained unchange. Therefore, it was concluded that bamboo could be more easily bended, because there were many parenchyma cells with thin-walled cells in the inner parts and more easily softened.

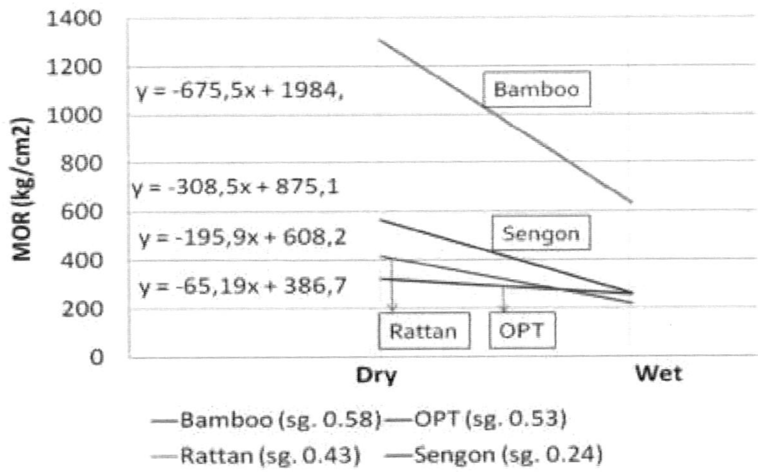

Figure 8. MOR of other lignocellulosic materials in softening conditions.

Wood and other lignocellulosic materials are composed of highly elongated cells whose walls have a complex multi-layered structure. In each layer, cellulose molecules are grouped together in long filaments called microfibrils embedded in a matrix composed of amorphous hemicelluloses and lignin. Both moisture and temperature act differently on the matrix and the microfibrils. Because of the crystalline nature of cellulose, water molecules cannot enter into the microfibrils. Bound water is held in the matrix and at the interface between the matrix and the microfibrils, and works as both a swelling agent and a plasticizer [20].

The elevation of temperature in wet condition softens the matrix, and its two main constituents, hemicelluloses and lignin, shift from the glassy state to something near the rubbery state. On the other hand, the microfibrils remain in the glassy state and are almost unaffected by moisture or heat. The softened wood and other lignocellulosic materials can be deformed easily. When a load is applied to the wood and other lignocellulosic materials, most of them are supported by the microfibrils at the local level. Softening of the matrix enables the relative displacement of the microfibrils so that their whole framework deforms elastically to adjust to the local loading. As lignin is a slightly cross-linked amorphous polymer, its deformation should be viewed as an increased viscoelastic strain rather than plastic flow [20].

When a deformed wet specimen is dried under restraint, the stress gradually decreases until it has disappeared. The departure of water molecules due to drying induces the reformation of hydrogen bonds between the molecules of the matrix. With the temperature decrease, this process leads to return the glassy state, where the elastic deformations of the micro fibrils and the matrix are frozen. The deformation is apparently fixed in the deformed state (drying set). However, as soon as the matrix is softened again through re-moistening and heating, most of the deformation is recovered due to the liberation of energy-elastic strain stored in the microfibrils and entropy-elastic molecular movements in the matrix [20].

It is well known that many amorphous polymers change from glassy state to rubbery state and are capable of plastic deformation (flow) in the case of linear polymers when heated above their glass transition temperature (or softening temperature). It was reported that the softening temperature of hemicelluloses and lignin of wood in wet condition were about –40°C and 60°C~80°C, respectively [20]. Above the softening temperature of the matrix, wood can be easily deformed to more than a strain of 50%. In this process, the stress remarkably decreases with time due to the micro-Brownian motions of molecules of the matrix. It is well known that the decrease of stress is accelerated by the departure of moisture during drying. Plastic deformation (flow) does not occur, because the matrix has a cross-linked structure.

Actually, most of the deformation was recovered by boiling after measurement. It is considered that the recovery of deformation was occurred by the liberation of energy-elastic strains stored in the micro fibrils and the entropy-elastic molecular movements in the matrix by boiling. This fact proved that the permanent fixation did not result from flow of the matrix. Previous researches on the mechanism of permanent fixation of radial compressive deformation of wood by heat or steam treatment could be found in [21–25].

4. Latest findings and future research

4.1. Development of furnitures with curved LVL

Research activities that have been conducted for years need to be applied and promoted to the users. In 2015-2017 through funding of Science and Techno Park (STP) Program, Theoretical and Technical Training, as well as Technology Transfer for Wood and LVL Bending Technique has been carried out to Small Wooden Furniture Industries at Sumedang District, West Java.

The product that can be easily developed is LVL curve, because it does not require a hot press machine. Solid wood bending is more difficult to apply and require substantial investment, as it requires a press machine equipped with heating components, and can only be made with a certain curved radius (Figure 3), when compared to LVL curve. The LVL curve requires only clamp tools made with a radius as expected and can be glued without a hot press (Figure 9). A small company, CV. Putra Lingga Jaya at Sumedang West Java has already produced these furnitures through License Cooperative Agreement.

Figure 9. Furnitures from curved LVL.

4.2. Development of bicycles with lignocellulosic material frames

From the above experiences, the technology of LVL curving has a potential to be developed as a new innovation of bicycle frame components [26]. Bicycle frame of LVL curve was previously designed in accordance with the expectation. From the design, a clamp was then made as a tool to form the bicycle frame.

Bicycle frames of LVL curve have been made from 2 mm thickness of rubber and teak wood veneers to form 20 mm thickness (10 layers). The wood veneers were wrapped around a clamp and glued with 300 gr/m^2 weight natural rubber latex and cured under room temperature conditions for 24 hours before they are formed into a bicycle frame, more clamps were used especially on the extreme curves. The process of making bicycle frame is shown in Figure 10.

Figure 10. Manufacturing of bicycle frame.

Since 2015, we have succeeded in making bicycle frames from LVL curve with various types and promoting them to some exhibitions. Two prototypes of fixie bikes were made with and without connections on their frames. Mountain bike types were made for both male and female, for which the upper part of female's frame was lower than male's frame.

It is expecting that the bicycles with LVL curve frame become an environmentally friendly vehicle for free-car areas, such as Botanical Gardens in the near future. By funding from STP Program, mass production process of the products will be developed through License Cooperative Agreement with a private company. In the near future, solid woods and other lignocellulosic materials, such as bamboo, rattan and oil palm trunk would also be developed for the bicycle frames.

Figure 11. Prototype of bicycles with LVL curve frames.

References

[1] ABC Four Corners 2016 *Background Information on Indonesia, Deforestation and Illegal Logging* (Wayback Machine).
[2] Matthews E (ed.) 2002 *The State of Forests Indonesia* (Bogor).
[3] Global Forest Resources Assessment 2010 *Trends in Extent of Forest 1990–2010* (Food and Agriculture Organization of the United Nations, Media Centre).
[4] John V 2014 *Rate of Deforestation in Indonesia Overtakes Brazil* (The Guardian).

[5] Dwianto W, Prianto A H, Amin Y 2004 Metode pemilihan jenis untuk kayu lengkung II–Pengujian tekan sejajar serat pada kondisi pelunakan *Prosiding Seminar Nasional VII MAPEKI (Makassar)* p A143–A152

[6] Darmawan T, Jayadi, Sudijono, Amin Y, Wahyuni I and Dwianto W 2005 Modifikasi alat pelengkung kayu skala pilot dengan menggunakan pemanas *J. Ilmu dan Teknologi Kayu Tropis* **4**(1): 1–8.

[7] Darmawan T, Dwianto W, Amin Y, Sudarmanto and Wahyuni I 2006 Pengaruh suhu dan waktu pemanasan terhadap tingkat fiksasi pelengkungan kayu Kepuk (*Sterculia* sp.) skala pemakaian *Prosiding Seminar Nasional IX MAPEKI (Banjarbaru).*

[8] Darmawan T, Dwianto W, Amin Y 2007 Fiksasi kayu lengkung dengan pemanasan oven *Prosiding Seminar Nasional X MAPEKI (Pontianak).*

[9] Darmawan T, Dwianto W, Amin Y, Prasetiyo K W and Subiyanto B 2009. Karakteristik LVL lengkung dengan proses kempa dingin *J. Ilmu dan Teknologi Kayu Tropis* **7**(2): 62–66.

[10] Darmawan T, Adi D S, Amin Y, Ismadi, Kusumah S S and Dwianto W 2016. Physical and mechanical properties of bent laminated veneer lumber (LVL) *Proceedings of the 7th International Symposium Indonesian Wood Research Society (Bandung)* p. 271–276.

[11] Dwianto W, Norimoto M, Morooka T, Tanaka F, Inoue M and Liu Y 1998 Radial compression of Sugi wood (*Cryptomeria japonica* D. Don). *Holz als Roh- und Werktoff* **56**: 403–411.

[12] Dwianto W, Morooka T and Norimoto M 1999 A method for measuring viscoelastic properties of wood under high-temperature and high-pressure steam conditions. *J. Wood Science* **45**(5): 373–377.

[13] Hamdan S, Dwianto W, Morooka T and Norimoto M 2000 Softening characteristics of wet wood under quasi-static loading *Holzforchung* **54**: 557–560.

[14] Hamdan S, Dwianto W, Morooka T and Norimoto M 2004 Fitting parameters for softening of wet wood under quasi static loading *Holzforschung* **58**: 134–137.

[15] Meyers and Chawla 1999 *Mechanical Behavior of Materials* p 98–103.

[16] McCrum, Buckley and Bucknell 2003 *Principles of Polymer Engineering* p 117–176.

[17] Baeurle S A, Hotta A and Gusev A A 2006 *Polymer* **47** 6243–6253.

[18] Sudijono, Dwianto W, Yusuf S, Iida I, Furuta Y and Minato K 2004 Characterization of major, unused and unvalued Indonesian wood species I. Dependencies of mechanical property on the changes of moisture content and/or temperature *J. Wood Science* **50**: 371–374.

[19] Iida I 1989 Changes of elastic and strength properties in the direction perpendicular to the grain by moisture content changes and by heating in water (in Japanese) *Mokuzai Gakkaishi* **35**: 875–881.

[20] Norimoto M and Gril J 1989 *J. Microwave Power and Electromagnetic Energy* **24**(4): 203–212.

[21] Dwianto W, Tanaka F, Inoue M and Norimoto M 1996 Crystallinity changes of wood by heat or steam treatment *Wood Research* **83**: 47–49.

[22] Dwianto W, Inoue M and Norimoto M 1997 Fixation of compressive deformation of wood by heat treatment *J. Japan Wood Research Society* **43**(4): 303–309.

[23] Dwianto W, Morooka T and Norimoto M 1998 The compressive stress relaxation of wood during heat treatment *J. Japan Wood Research Society* **44**(6): 403–409.

[24] Dwianto W, Morooka T, Norimoto M and Kitajima T 1999 Stress relaxation of Sugi (*Cryptomeria japonica* D. Don) in radial compression under high-temperature steam *Holzforschung* **53**: 541–546.

[25] Dwianto W, Morooka T and Norimoto M 2000 Compressive creep of wood under high-temperature steam *Holzforchung* **54**: 104–108.

[26] Dwianto W, Darmawan T, Jayadi and Adiana D D 2017 Development of bicycles with LVL curve frames *Proceeding of International Symposium on Bioeconomics of Natural Resources Utilization (Bogor)* p 586–591.

IOP Conf. Series: Earth and Environmental Science **359** (2019) 012014 doi:10.1088/1755-1315/359/1/012014

Characteristics of Particleboards Made from Agricultural Wastes

L. Astari[*], Sudarmanto, F. Akbar

Research Center for Biomaterials, Indonesian Institute of Sciences,
Cibinong Science Center, Cibinong-Bogor 16911, Indonesia

[*]email: astari@biomaterial.lipi.go.id

Abstract. The growth of particleboard production is parallel with the growth of world population. Yet, supply on wood chips as raw material for particleboard facing the shortage. Indonesia as a tropical country has various source of renewable resources, including agricultural residuals and non-wood forest product. This research was conducted to investigate the properties of particleboard made from core and rattan bark, from corn stalk and wood waste. Particles with moisture content less than 5% were mixed with adhesive. Phenol formaldehyde (PF) 10% was used as the adhesive. Mixed particles were then hot pressed for 10 minutes at 150°C with 25 Kgf/cm^2. Targeted density of the board was 0.7 g/cm^3. Prior to testing, boards were conditioned for seven days at room temperature. Testing was conducted for physical and mechanical properties as well as the surface roughness. Standard for physical and mechanical properties reffered to the JIS A 5908-2003. Results show that the particleboard from corn stalk had the lowest thickness swelling which was 16.7%. The highest modulus of rupture was given by particleboard from wood waste which was 21.69 MPa.

1. Introduction

Wood-based panels includes particleboard are getting more important since they can substitute solid wood for furniture and construction sectors. Particleboard is still the most excellent wood-based products due to its advantage, such as evenly distributed mechanical properties in all part of the board. Furthermore, particleboard can be produced from less-desired wood and several lignocellulosic non-wood materials such as bamboo and certain type of agricultural wastes, such as wheat straw, rice straw, sorghum stalks, coir, corn cob and corn stalks. The various raw materials for particleboards results in high yield production compare to those of solid wood. Specific application of particleboard is quite extensive such as buildings, automotive interiors, packaging and furniture [1]. All those applications are broadly used in Indonesia.

Data from FAOSTAT mentioned that in 2017, particleboard production in Indonesia was 125,000 m^3. This makes Indonesia in 48 of world rank on particleboard production which was 0.1% of world share. In addition, during the last three years, the global production of particleboard is escalated. In 2017, the worldwide production of particleboard was approximately 95.5 million m^2 [2]. Demand on particleboard forecasted to rose due to the increase on necessity in furniture and building materials. The high demand on both interior and exterior housing parts as a result of high rate on population is increasing, especially in Indonesia which populated with approximately 266,8 million people in 2018.

Therefore, the objectives of this research were to investigate the suitability of several agricultural wastes as raw material for particleboard as well as the properties of particleboard.

2. Material and method

Material used for this research were core and skin of rattan waste, corn stalk, coir and wood waste. Rattan waste was supplied by PT. Rimba Makmur Utama (RMU), Katingan. Corn stalk and wood waste were obtained from traditional farm in Dramaga, Bogor. Phenol formaldehyde as adhesive was purchased from PT. PAI Probolinggo.

2.1 Particle Preparation

Raw materials were manually chopped using ring flaker machine to obtain small size particle. Particle were sun-dried to reduce the moisture content. Following step was sieving the particle to retain 4-14 mesh. Particle at this stage has moisture content above 25%, thus oven dried process should be carried out. Oven drying was successfully reduce the moisture content particle into less than 5%, so that the particle can be processed into particleboard. Dry particle and adhesive were weighing according to formula. Formula calculation depends on targeted density of the particleboard.

2.2 Particleboard Production

Targeted density of particleboards was 0.7 g/cm^3 board and the board dimension was 30 cm x 30 cm x 1 cm. The phenol formaldehyde (PF) content was 10%. Particle and adhesive were mixed using drum mixer and spray gun. Afterwards, the mixture of particle and adhesive was spread in a 30 x 30 wood-square mould. Hot pressing was conducted for 10 minutes at temperature 150°C with pressure 25 kgf/cm^2. The particleboard was conditioned for 7 days at room temperature prior to physical and mechanical testing.

2.3 Particleboard Testing

2.3.1 Physical Properties. Testing for physical properties includes particleboard density, moisture content, thickness swelling and water absorption after 24 hours immersion.

- Particleboard density

  Density was calculated according to the following equation:

$$D = \frac{m}{v} \qquad (1)$$

D = density (g/cm^3)
m = mass (g)
v = volume (cm^3)

- Moisture content

  Dimension of test specimen for moisture content was 5 x 5 cm. The equation for moisture content is as follows:

$$mc = \frac{w1-w2}{w2} \times 100 \qquad (2)$$

mc = Moisture content (%)
w1 = initial weight of the samples (g)
w2 = weight after oven (g)

- Thickness swelling and water absorption

  Basically, the investigation on thickness swelling is based on the dimensional change of the test specimen (5 x 5 cm). The test specimen was immersed in water for 24 hours.

$$TS = \frac{T0-T1}{T1} \times 100 \qquad (3)$$

TS =Thickness Swelling (%)
T0 = Wet thickness after water immersion (mm)
T1 = Dried Thickness (mm)

- Water absorption was calculated with the following equation:

$$wa = \frac{w2 - w1}{w1} \times 100 \qquad (4)$$

wa = water absorption (%)
w1 = Weight at dry condition (g)
w2 = Weight after immersion (g)

2.3.2 Mechanical Properties. Mechanical properties that were tested were flexural strength (modulus of rupture/MOR and modulus of elasticity/MOE), internal bond and screw withdrawal. Flexural strength was tested using three points static bending test using Shimadzu Universal Testing Machine (UTM) 50 kN load cell. Internal bond and screw withdrawal were also tested using the same machine with different cross head.

2.3.3. Surface Roughness. Mitutoyo surface roughness tester type SJ-210 was used to test the average of surface roughness (Ra) of the particleboard. Measuring speed and pin top angle were 0.02 in/s and 90°.

3. Results and discussion

3.1 Physical Properties

3.1.1. Density and moisture content. The density of particleboard was ranged between 0.78 – 0.73 g/cm^3, this result indicates that all the particleboard reached the targeted density (0.7 g/cm^3). Particleboard made from wood waste has the highest density (0.78 g/cm^3). Figure shows that all the particleboards have density that fulfil the Japan Industrial Standard (JIS) A 5908:2003 [3]. The standard requires the density of particleboard between 0.4-0.9 g/cm^3. Density of particleboard is greatly affected by the density of raw materials. The graph shows that particleboard made from wood particles has the highest density compare to four other boards that made from non-wood materials. Those different value of board density as a result that wood particle has higher density compare to those corn, coir and rattan.

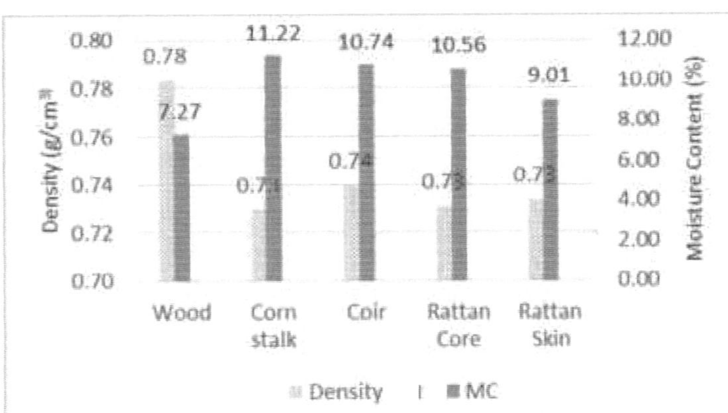

Figure 1. Density and moisture content of particleboard.

Moisture content of particleboard was 7.27-11.22%. The acceptable range of moisture content based on JIS A 5908:2003 is between 5-13%. Thus, all the particleboards have acceptable moisture content based on the standard. Particleboard from corn stalk had the highest moisture content (11.2%), this was due to the characteristics of corn stalk that comprises by a pithy core and the long fibres at its outer layer [4]. Those components remarkably hydrophilic that easily absorbs moisture. Moisture content of raw materials is not totally released during the hot pressing therefore the moisture content of raw materials directly affected the moisture content of particleboard. Furthermore, liquid adhesive such as phenol formaldehyde contains water that can increase the moisture content of particleboard around 4-6% [5].

3.1.2. Thickness Swelling and Water Absorption. JIS standard mentioned that acceptable range of thickness swelling for particleboard is maximum 12%. The graph in Figure 2 shows that the best thickness swelling was performed by particleboard from corn stalk with the thickness swelling 16.7%, while the unsatisfied percentage of thickness swelling given by particleboard using coir which was 37.10%. Therefore, all the boards far above the standard value.

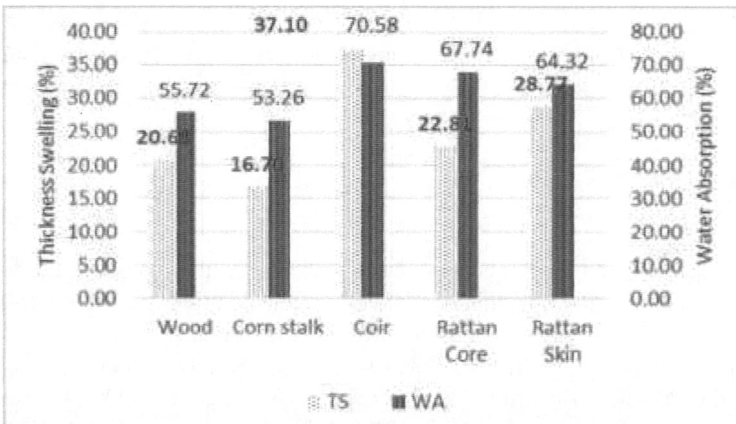

Figure 2. Thickness swelling and water absorption of particleboard.

Several factors that might be contributed to this condition are percentage of adhesive and the insufficient time for transformation of hydrophilic component of the fibres into hydrophobic products. it means that the addition on pressing time predicted to increase the thickness swelling propertie [6]. Water absorption value is inline with thickness swelling percentage. As it is seen from the graph, the higher percentage of thickness swelling also the more particleboard absorbs water. There is no certain value on JIS regarding the percentage of water absorption for particleboard. The application of water repellent such as wax or paraffin may decrease the water uptake which leads to better thickness swelling properties.

3.2. Mechanical properties

3.2.1. Flexural Strength. The results test for MOR of all the particleboards are depicted in Figure 3. MOR values were around 10.1-21.7 MPa. The lowest MOR performed by particleboard from corn stalk and the highest given by particleboard from wood particles. JIS standard categorized particleboards into three types according to its value of MOR, type 8 with minimum MOR value 8 MPa, type 13 and 18. The result suggests that only particleboard made with corn stalk was classified in type 8. Another four particleboard are categorized as particleboard type 18.

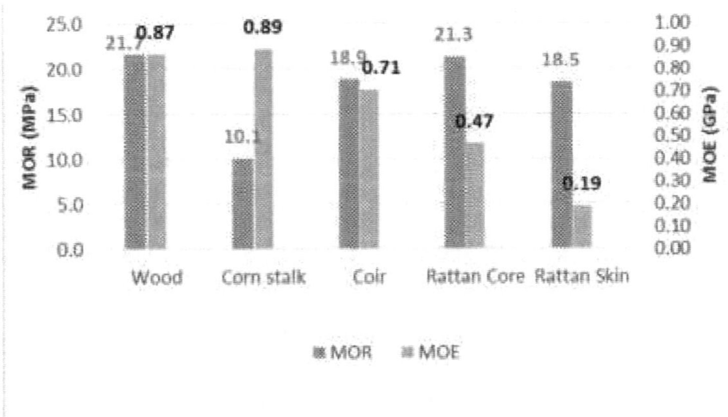

Figure 3. Flexural strength value of particleboard.

Particleboard made from rattan core has MOR value that slightly different with particleboard made from wood. The low value of MOR on particleboard using corn stalk was caused by the small surface contact between particle and adhesive. Furthermore, corn particle is very light compared to the other materials. Beside percentage of adhesive, particle geometry and particle density has major impact on MOE and MOR of particleboard [7]. Regarding adhesive percentage, the more adhesive gives the higher MOR, yet increase the production cost. High amount of adhesive increases the bond contact among individual particles which led to a superior MOR [8].

3.2.2. Internal Bond. Internal bond values of particleboards were 0.35-1.46 MPa. Particleboard from wood particles gave the topmost value while particleboard from rattan core was the lowest. Since JIS A 5908:2003 required 0.15 MPa as the minimum value of internal bond for particleboard type 8 thus all particleboards satisfy the standard. Distribution of adhesive on particle significantly influenced internal bond properties.

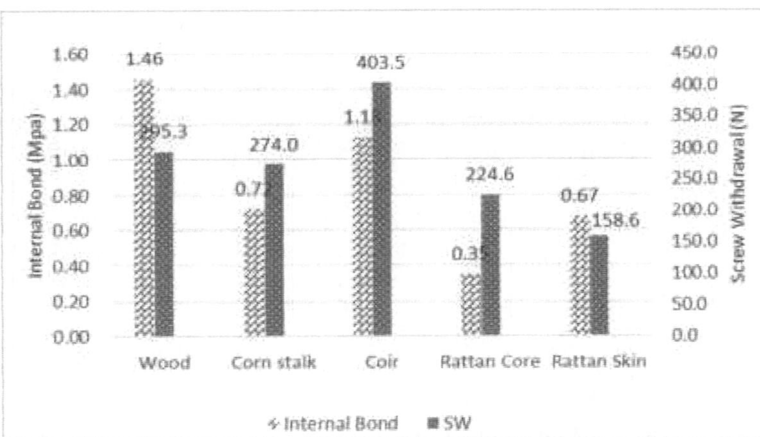

Figure 4. Internal Bond and Screw Withdrawal value of Particleboard

Internal bond of panels is the most depictive mechanical properties for characterising particleboard, both for internal and external laboratories purposes. Using the result from internal bond testing, particleboard can be promptly assessed for suitability for edge machining, among other properties [9]. According to those insight, particleboard made from wood waste particle has an early sign for superior mechanical properties.

3.2.3. Screw Withdrawal. The highest screw withdrawal value on this research was 403.5 N obtained from particleboard made from coir. Since the JIS require minimum 300 N for screw withdrawal, particleboard with coir give the only screw withdrawal value that meet the JIS standard. Particleboard made from wood waste particle was slightly under the standard which was 295.3 N, the lowest value was given by particleboard made from rattan skin which was158.6 N. The high value of screw withdrawal is influenced type of particle as its raw materials [10].

3.2.4. Surface Roughness. The average surface roughness of particleboard in this research was ranged 3.34-6.09 μm. Particleboard from wood had the smoothest surface while particleboard from rattan had the roughest. According to Hiziroglu and Suzuki, acceptable range of average surface roughness for particleboard was 3.67-5.56 μm [11]. Therefore, only particleboard from wood particles and rattan skin were acceptable according to the reference.

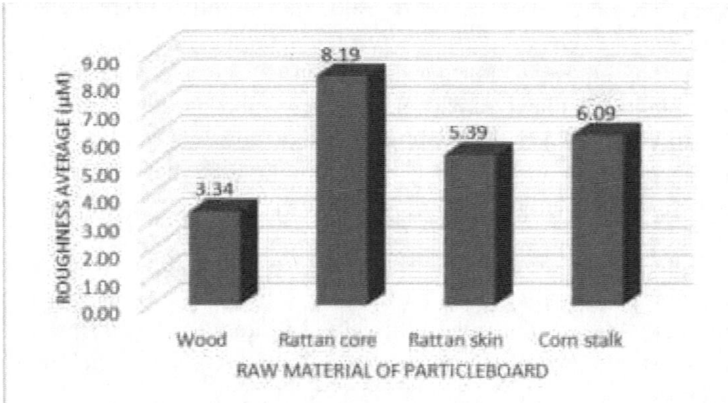

Figure 5. Roughness average of particleboards.

Surface quality of particleboard and other wood-based products play a major role on finishing and bonding characteristics. Data of surface characteristics on this research can be useful to estimate amount of finishing material that will be used There are several ways to evaluate the surface characteristics, stylus method which is used in this research is argued as the simplest way yet provide reliable data. [12]. The better characteristics of surface roughness on particleboard from wood were also due to the better interaction of wood particles with PF as adhesive. Another finding reported that particleboard from non-wood materials (wheat straw) with urea formaldehyde resin as adhesive gives superior surface roughness [13]. Thus, surface roughness value also determined by the suitability of particles type with the adhesive.

4. Conclusion

Agricultural wastes that was used in this research were suitable as raw material for particleboard. All the boards were meet the targeted density which was 0.7 g/cm^3. The highest moisture content was performed by particleboard made from corn stalk (11.2%) and the lowest was particleboard made from wood (7.27%). Thickness swelling of all particleboards had not meet the standard, yet, the best was 16% given by particleboard made from corn stalk. The highest thickness swelling was 37.10% given particleboard using coir, consequence with water absorption value which around 70%. Modulus of rupture (MOR) and internal bond (IB) of particleboard satisfied the standard of JIS A 5908-2003. Particleboard from corn stalk hasd the roughest surface with roughness average value approximately 8.19 μm, yet it had better physical properties among others.

Acknowledgement
Authors acknowledge Research Center for Biomaterials (DIPA Program), SATREPS and JASTIP Program for research funding.

References
[1] K. Nakano, K. Ando, M. Takigawa, and N. Hattori, "Life cycle assessment of wood-based boards produced in Japan and impact of formaldehyde emissions during the use stage," *Int. J. Life Cycle Assess.*, vol. 23, no. 4, pp. 957–969, 2018.
[2] FAOSTAT, "Indonesia: Particle board, production volume (cu. meters)," 2019. [Online]. Available: http://www.factfish.com/statistic-country/indonesia/particle board%2C production volume. [Accessed: 04-Mar-2019].
[3] Japanese Standard Association, *Japan Industrial Standard (JIS) A 5908:2003*. 2003.
[4] R. Jarabo, M. C. Monte, E. Fuente, S. F. Santos, and C. Negro, "Corn stalk from agricultural residue used as reinforcement fiber in fiber-cement production," *Ind. Crops Prod.*, vol. 43, no. 1, pp. 832–839, 2013.
[5] J. . Bowyer, R. Shmulsky, and J. . Haygreen, *Forest Products and Wood Science*, 5th ed. Blackwell Publishing, 2007.
[6] Z. Gao, X. Wang, H. Wan, and G. Brunette, "Binderless panels made with black spruce," *BioResources*, vol. 6, no. 4, pp. 3960–3972, 2011.
[7] L. Sekaluvu, P. Tumutegyereize, and N. Kiggundu, "Investigation of Factors Affecting the Production and Properties of Maize Cob-Particleboards," *Waste and Biomass Valorization*, vol. 5, pp. 27–32, 2014.
[8] B. Ajaye, "Durability characteristics of cement-bonded particleboards manufactured from maize stalk residue," *J Res*, vol. 22, no. 1, pp. 111–115, 2011.
[9] P. D. Palacios, F. G. Fernandes, A. Garcia-Iurela, B. Gonzales-Rodrigo, and L. G. Esteban, "Study of the influence of the physical properties of particleboard type P2 on the internal bond of panels using artificial neural networks," *Comput. Electron. Agric.*, vol. 155, pp. 142–149, 2018.
[10] R. M. Rowell, R. A. Young, and J. K. Rowell, *Wood Chemistry and Wood Composites*. CRC Press, 2005.
[11] S. Hiziroglu and S. Suzuki, "Evaluation of surface roughness of commercially manufactured particleboard and medium density fiberboard in Japan," *J. Mater. Process. Technol.*, vol. 184, pp. 436–440, 2007.
[12] Z. W. Zhong, S. Hiziroglu, and C. T. M. Chan, "Measurement of the surface roughness of wood based materials used in furniture manufacture," *Measurement*, vol. 46, no. 4, pp. 1482–1487, 2013.
[13] S. M. Hafezi and K. D. Hosseini, "Surface characteristics and physical properties of wheat straw particleboard with UF resin," *J. Indian Acad. Wood Sci.*, vol. 11, no. 2, pp. 168–173, 2014.

AUTHOR INDEX

Institute of Physics
Dirac House, Temple Back
Bristol BS1 6BE UK

ISSN: 1755-1307
ISBN 978-1-7138-0155-9

CURRAN ASSOCIATES INC.
proceedings.com

International Conference on Materials Technology and Energy (ICMTE 2017)

IOP Conference Series: Materials Science and Engineering
Volume 217

Miri, Malaysia
20 – 21 April 2017

IOP
Institute of Physics